Cable Advertiser's Handbook

Second Edition
First edition titled
CABLE: An Advertiser's
Guide to the New Electronic
Media

Ronald B. Kaatz

CRAIN BOOKS, an imprint of
NATIONAL TEXTBOOK COMPANY • Lincolnwood, Illinois U.S.A. 

Published by Crain Books, an imprint of National Textbook Company, 4255 West Touhy Avenue, Lincolnwood, Illinois 60646-1975

Manufactured in the United States of America.

5 6 7 8 9 0 ML 9 8 7 6 5 4 3 2 1

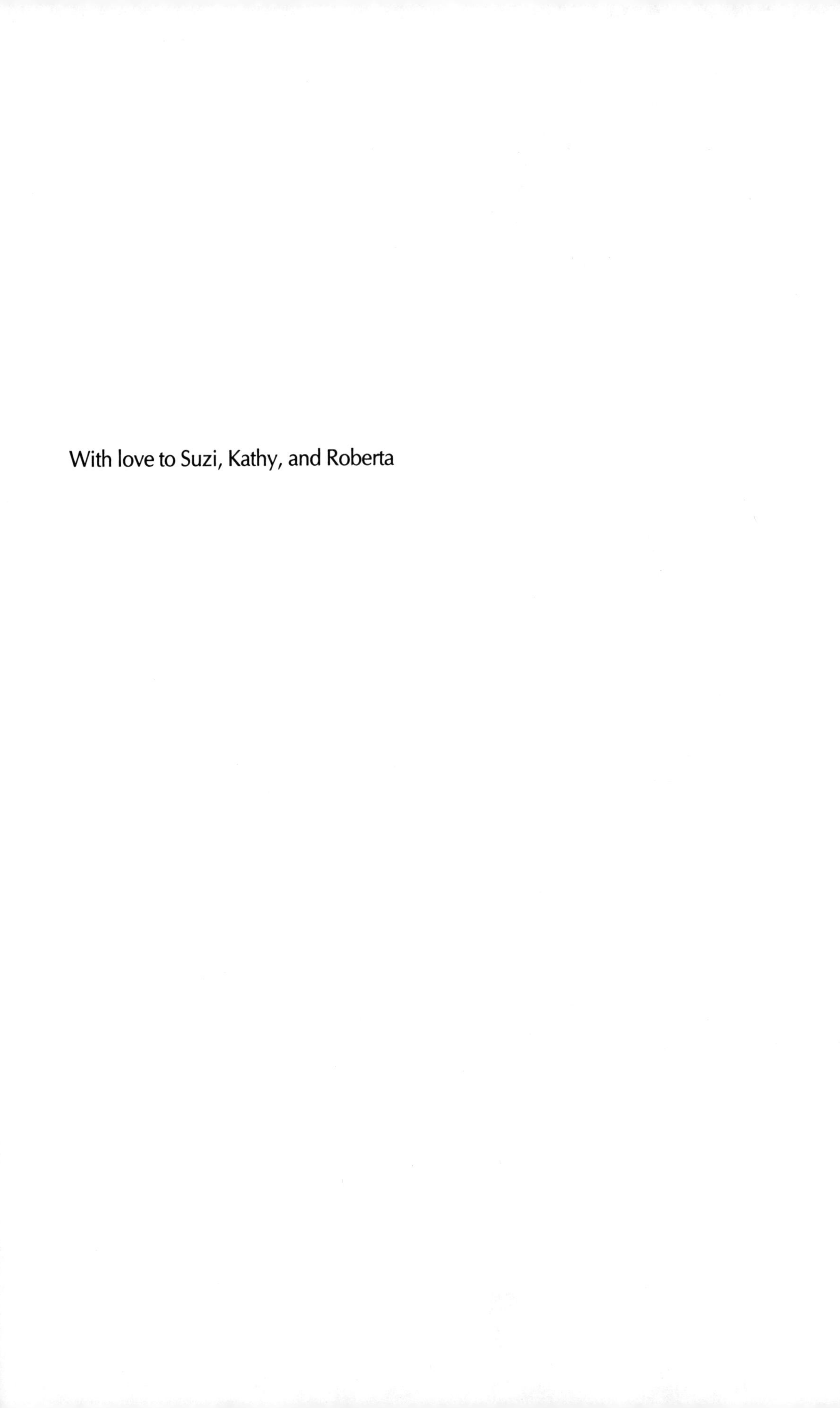

With love to Suzi, Kathy, and Roberta

Contents

Figures

Tables

Preface

If someone had suggested in 1975 that a book be written on advertising and video technology, the immediate response would have been, "But what will you say after the first couple of pages?"

And yet today, more is written by more reporters in more periodicals about Videotech™ (or what is more generally called the New Electronic Media) than has been written about any other advertising or marketing subject in decades. An entire new lexicon of terminology has been created—concepts and words that either didn't exist or were discussed only in the most technical circles a few years back. A dozen or more new magazines have sprung up to deliver Videotech™ news and features to the communications industry, advertisers, and viewers. And older publications have created special new media sections and departments. Hardly a week goes by without a meeting, seminar, conference, or convention devoted to cable television or home video. These events all underlie the growing importance of the new video technology to everyone involved in the dissemination of information and the marketing of goods and services.

In the years following World War II, no two developments more significantly shaped the world of advertising than television and the computer. With television, marketers at last had available to them a medium that could bring their products to life and present them instantly to millions of consumers. And, later, the computer gave these same marketers the ability to process and analyze millions of pieces of information on the media habits, demographic characteristics, and purchasing patterns of these prospective customers.

The "Third Wave" of progress is now upon us. It promises to have as great an impact on everyone involved in the production, distribution, and sale of goods and services, as have television and the computer.

Television provided advertisers with the means of reaching millions of people at a single point in time in an entertainment environment. The new video technology allows these same advertisers to zero in more closely on thousands of their best customers in an environment that is more finely tuned to their product or service. At the same time, the new media opens up the world of video to advertisers whose narrow marketing targets, confined geographic sales areas or limited budgets precluded them from television.

Actually, the new electronic media represent more than a single medium. They encompass an entire spectrum of means to deliver entertainment, information, and advertising into our homes via the television set. These include cable television, video information services, and home video players and recorders.

The arrival of these new media has created not only new opportunities, but many problems for many of the companies involved.

Since the first edition of *Cable*, as it was titled, was published in September 1982, there has been a busy period of mergers, contractions, and disappearances within the industry. It began with the announcement in September 1982 that CBS Cable would shut down after losing $30 million. Then, in early 1983, The Entertainment Channel followed suit after only nine months of service and a loss of some $35 million. In late 1984, Turner Broadcasting faced the music and shut down their Cable Music Channel, only four weeks after it was launched to compete with MTV. In the merger category, ABC Video Enterprises and Hearst Corp's women's network Daytime merged with Viacom's Cable Health Network to form Lifetime, a new 24-hour-service. At the same time, ABC Video's cultural service ARTS merged with RCA's Entertainment

Channel to give birth to Arts & Entertainment. Ted Turner proved that there was not room for a third 24-hour news service. He purchased Satellite News Channels, folded it, and used its subscriber base to enlarge those of Cable News Network and CNN Headline News. The STV and MDS industries have shrunk dramatically during the last several years, with many systems shutting down completely and others losing large numbers of subscribers. According to industry analyst, Paul Kagan, the number of STV and MDS subscribers plunged from 2.1 million in 1981 to 900,000 in 1984 and are expected to fall to 500,000 by 1990. Fierce competition in the video game and home computer industry resulted in drastic price cutting of both hardware and software, and enormous declines in profits that were reflected in significant stock price slides.

One reason for the game industry's problems has been the boom in home computer sales as consumers discovered they could be used to play games *and* do much more. By early 1985, home computers were already in more than 10 percent of the nation's homes. Their strong sales, however, were no guarantee of profits and competition forced rapid changes in product design and deep price discounts. Texas Instruments, Mattel, Timex and Coleco all dropped out of the low-priced end of the home-computer market as the public began to show a growing preference for the more sophisticated, expensive models.

The videodisc industry received a severe blow in April 1984 when RCA, the leader in the field, announced it would stop manufacturing and marketing videodisc players. However, in another home electronics area, the marketplace was far brighter. Demand for VCRs hit an all-time high, and by early 1985, one out of five homes were both watching prerecorded tapes and busily "time-shifting." It is now forecast that by the end of the decade, VCRs will be in half of the nation's homes.

Whenever an announcement is made of a shake-up or shake-out within the cable or home electronic field, the news media immediately begin to ask, "Is this the start of a trend?" "Will there be more?" "Is the bloom off the rose?" What they seem to forget is that:

- Every year, thousands of new products are introduced in stores across the country. The vast majority of these fail to capture either the consumer's fancy or pocketbook.

- Broadway backers, time after time, invest in new shows they are convinced will make them rich. In most cases, all they ever get out of it are a couple of tickets to the press preview followed by a cocktail-buffet.
- Every fall, a new network television season gets underway. By the next year, less than a handful of all the shows that premiered will return to the screen.

and

- The enormous growth which has taken place in the new media arena has all done so in a very short period of time—since the start of the 1980s.

In reality, it is inevitable that over the coming months and years that:

- Competition in the home video equipment field will heighten.
- There will be additional cable programming failures.
- Cable systems being built or franchised today will fail to turn in the profits that were expected.
- Electronic information services for the home may develop far more slowly than was once expected.
- Mergers and co-ventures now existing or on the drawing boards will fall apart or fail to materialize.

The winners in all areas will be those services or systems that realistically appraise their revenue potential and are able to keep both programming and operating costs in line with it.

What is happening is that the new media are entering a period of new maturity—of shake-ups—and of settling in place.

To assume the world five years hence or even a year from now will look exactly like it does at present is sheer foolishness. Those services that offer strong viewer and advertiser values will succeed just as do those television shows, radio services, magazines, and newspapers that fulfill their stated goals. Others will fall by the side of the road.

One added note: If television were in its infancy today, and it was announced that "Dozens of shows will premier this year and only a fraction of these will be around next season," all of the critics would probably say: "We told you TV wouldn't make it!"

And they'd be wrong!

This new edition, now titled *Cable Advertiser's Handbook*, focuses mainly on cable television, the most developed of the new media, from the standpoint of advertising potential. At the same time, it offers suggestions as to how the other new media services also can be used to communicate advertising and information in an effective manner. And, of course, as it focuses on the opportunities, it will also help you solve some of the problems you will encounter as you begin to travel down the new media advertisng road.

In a nutshell, this book is targeted at everyone involved in planning, buying, selling, and developing advertising for—or just learning about—cable and other new media. It is an "idea starter" and is aimed at getting you to start thinking about how to use the new media to communicate in new and creative, yet practical and down-to-earth ways. This second edition reflects all those changes that have taken place since the first edition was published.

Gorrell, 1982. *The Charlotte News.* Courtesy of Field Newspaper Syndicate.

CHAPTER 1

The Early Days

The new media of the 1980s encompass a veritable gourmet menu of viewing options and video gadgetry. Actually, there are so many options that many homeowners are converting what once was a family room or den into a home video center.

The latest in consumer electronics marketing is the concept of component TV. Instead of buying an "all-in-one" television set, the consumer buys the video monitor, receiver, and speakers separately. Once the "basics" are installed, he can later attach home computers, video games, VCRs and video disc players, teletext, home-security systems as well as stereo systems. While expensive, component TV provides the basis for setting up the ultimate home entertainment and information center and, actually is similar to the way in which audio products began to develop in the 1960s. Everything would be controlled from a single source selector allowing for the instantaneous delivery of entertainment and information complete with stereo sound.

And being "wired for pictures and sound" is only the beginning. Already in the little town of Benecia, California (just 32 miles north of San Francisco), we have the first "computer-ready" community. Homes in a new subdivision are equipped with a prewired family room and bedroom closet large enough to house a computer, printer, and other accessories. And if you don't already have any of these items, the developer will let you tack $3,500 on to your mortgage to acquire them.

A Century of Developments: the 1880s to 1980s

It may surprise you, but as we focus on advertising and the new media of the 1980s, we really go all the way back to the 1880s and the earliest beginnings of television. It was then that a patent was taken out on a rotating disc that had spiral perforations through which it could scan a scene. Unfortunately, nothing was really done with this device until half a century later. Then, during the 1920s and 1930s, an Englishman named John Baird developed a television set that people could buy in London department stores. Baird broadcast programs using an improved version of the old rotating disc and even figured out a way to record his images on what had to be the world's first "video disc." Unfortunately, neither the discs nor the "Televisor" set itself bore much resemblance to today's TV sets. More importantly, no one was either ready or very much interested in developing advertising for them.

The real beginnings of television, as we know it, probably took place in San Francisco, when Philo Farnsworth first demonstrated *color* television in 1927. A key role was then played by Vladimir Zworykin, who invented the first *electronic* TV camera pickup tube during the 1930s. This brought television out of the labs and into the spotlight where it was demonstrated to mass audiences at New York's 1939 World's Fair.

World War II held up the production of sets and the growth of stations, but once the fighting ended, the new medium took off. Sets were coming off the assembly lines, and stations were growing throughout the country. All that was missing were programs to fill the time. Very quickly, however, the networks came along and not only did this, but, in the process set in motion the creation of the most powerful advertising device that had yet been imagined—the television commercial.

The Origins of Cable

Why CATV Came About: Reception and Variety

Television had barely gotten underway in the late 1940s when the first new media seeds began to sprout. It was back in 1948 that the earliest cable systems were born in remote areas of Pennsylvania and Oregon. Known then as Community Antenna Television (CATV), its function was simply to bring TV signals into communities where off-air reception was either nonexistent or very bad because of distance or interference from mountains. The earliest systems were often started by appliance dealers so they would be able to sell television sets to the folks in their community.

In Astoria, Oregon (5,000 homes), it is reported that Leroy Edward Parsons wired the first system with an initial subscriber list of three households. The first commercial CATV system was installed two years later in Lansford, Pennsylvania, after which they began springing up all over the map. To bring in the signals, CATV systems employed master antennas that were mounted on tall towers erected on the plains or on mountain tops. The cable system operator then delivered the signals to subscribing homes in the community via coaxial cable that was either strung overhead on utility poles or underground in conduits.

In small communities without any television service, CATV provided a brand new medium. And, in many other towns that had only one or two television stations, viewers now had the full program offerings of three national networks. Cable even helped reception in the cities—in San Diego, which is encircled by mountains, and in Manhattan with its giant steel buildings. The term *subscriber* was aptly applied to those homes that were hooked up to cable. They paid an installation charge to be connected, and then they paid a monthly fee of $3 to $5 just as if they were subscribing to a local newspaper.

In the 1950s, the system operator's master antenna was joined by microwave relay equipment that extended the range of originating station pickup. Common carrier terrestrial microwave routes emanated from major market areas and transmitted the network signals to cable systems hundreds of miles away. Early on, it was recognized that a function of CATV could be to supply channel diversification. Suppliers like Jerrold Corporation developed sophisticated equipment that permitted reception of 6 to 12 channels.

While these advances were well received by viewers, many television station operators (network affiliates) viewed diversification with alarm. In the early 1960s, programming was "bicycled" to many smaller market stations by film (and later tape). If a cable system could import a network program weeks in advance of a local network affiliate's own scheduled airdate, the local station's own audience delivery could be affected seriously. And, of course, the emergence of the independent television station in the larger metropolitan markets—and its subsequent microwaving to the small communities—further affected a local station's ratings and, hence, what could be charged for advertising time.

Early Rules and Regulations

The concerned voices of station owners were heard very loud and clear in Washington, and in 1964 the Federal Communications Commission issued its first cable regulation. Cable operators were required to "black out" programming that came in from distant markets and duplicated a local market station's own programming, if the local station demanded it. Just as a point of reference, there were only about one million cable homes in the entire country at the time this took place! (It is interesting to note that even in the early days of cable, "audience fragmentation" was a factor that stations had to worry about, even though they didn't know it by that term.)

Despite the imposition of FCC regulations, cable systems continued to grow throughout the country. They served a much-wanted function of bringing program diversity to areas of limited program availability. The broadcasting industry, however, was still alarmed by the potential of cable, and in 1968 the FCC, for all practical purposes, imposed a "freeze" on its further growth. The new ruling said that if a cable system was going to be located within 35 miles of a television market, it could offer its subscribers only those TV signals that were available off air (i.e., not broadcast). This served to bring cable development in the larger markets to a virtual standstill. To a certain extent, cable was once again relegated to fringe areas and markets with poor off-air reception. As another benchmark, there were fewer than three million cable homes at this time. This represented about 5 percent of the television homes in the country.

Who Owned the Systems: Investors and Finance

In its earliest days, CATV systems were generally owned singly by local business people. They were small—in 1952 there were 70 CATV systems with a total of 14,000 subscribers (or 200 subscribers per system). It was not unusual for a local hardware store or electrical contractor to own a system. After all, he had the materials, equipment, and know-how to build it. This image of CATV as a "small business community service" was soon altered as large groups quickly realized the profit potential in the industry. Because of popular demand, liberal depreciation policies, and debt-leveraged ownership, many early systems showed rates of return on invested capital exceeding 15 percent, while returns as high as 50 percent were claimed by a few. And this was before the days of the added pay services! It is no wonder that cable was quickly discovered by many of the largest names in the communications industry. In contrast to its early "Ma and Pa" status, the cable industry today is dominated by a number of major companies known as MSO's (Multisystem Operators). The Top 15 cable MSO's today account for about half of the industry's total subscribers.

The Early Days of Cable Advertising

With only 5 percent television home penetration, CATV systems still had substantial influence on the broadcasting industry as the 1960s came to an end:

- Most apprehension of CATV expressed by the networks was oriented to the future. Market fragmentation and possible audience loss and, hence, financial injury to their local affiliates began to concern them. There was even concern that a "fourth network" of CATV systems might be formed to compete in the advertising marketplace.

- To many broadcasters, the growth of CATV posed a large threat to the survival of many local stations, especially the UHF stations, which were the most vulnerable to competition. In a 1968 decision, the FCC ruled in favor of the local stations and restricted the expansion of CATV in San Diego. (Even so, in late 1984 Cox Cable in San Diego was the largest system in the country.)

- The subject of copyright was rearing its head as CATV operators took broadcast programs off the air without permission of the music and program copyright holders. At the same time program costs to broadcast stations whose audiences were being boosted by cable were increased.

Cable's impact on advertising in the late 1960s was not yet very significant, but many marketers were already looking ahead to the problems and opportunities that would be created:

- Advertisers foresaw possible audience losses to their spot buys that would result from the viewing of distant stations by CATV subscribers in the market.
- On the other hand, advertisers might gain "bonus" coverage through transmission of a station's signal from one market to another. Unfortunately, such coverage might not be desirable. For example, a McDonald's or Burger King promotion in Denver might be carried by CATV into western Nebraska where the item would not be available.
- CATV could create a new, inexpensive local advertising medium if cable operators sold commercial time to help pay for cable-originated programming.
- While CATV had little effect on most advertising agency media buys, any future reshuffling of markets and market fragmentation would seriously confuse the time-buying function.
- The problems of audience measurement were not yet major, but it was recognized that they would become significant for the rating services as CATV spread. There were already certain smaller markets (Salisbury, Santa Barbara, Clarksburg-Weston, Marquette, and Missoula) where CATV penetration was over 40 percent and where stations were beginning to feel their audiences were not accurately reflected in the ratings.

Probably the greatest use of cable by advertisers in these early days was for advertising testing. It started early in 1964 when the Center for Research in Marketing in Peekskill, New York, contracted with a Port Jervis, New York, CATV firm to use its commercial testing facilities. The project was possible because of a "split cable." The basic concept of the test system was fairly simple. Test commercials were slipped into regular network programs as the shows were fed via cable into subscriber homes. The subscribers were split into two separate but demographically matched samples. The researchers varied copy, media weight, commercial length, frequency, etc., and studied the effect of all this on the two samples of viewers. Unfortunately, because the supermarket scanning devices now in use in such tests had not yet been invented, the precision of these tests was limited.

By the end of the early days of cable, the industry had 2,260 systems and 3.6 million subscribers. Overall, 1 out of 16 homes in the country was cable equipped, with the majority of these homes located in smaller towns and rural communities. (See Table 1-1.)

TABLE 1-1 Cable Penetration, 1969

Nielsen Market Areas	Percent CATV Penetration of TV Homes
Major Metro Areas	1.6
Medium Sized Cities	4.8
Smaller Towns	16.8
Rural, Farm Areas	10.2
Total United States	6.4

Source: A.C. Nielsen, November 1969.

More Rules and Regulations

As the 1960s drew to a close, the big question concerning CATV was, "What might exist tomorrow?" While the outlook for CATV seemed favorable, its future was clouded by just what direction FCC regulatory proposals would take. Three possible scenarios for the future existed:

- If a series of restrictive laws or decisions were produced, CATV could be expected to do no more than serve those areas of marginal reception it was originally created to serve.
- If regulatory decisions were favorable *and* if there was a real demand for a broader choice of programming and not just clearer reception, CATV would expand into the major urban markets.
- Finally, if favorable regulatory decisions and market demand were joined by increased expenditures by subscribers and system operators, the services offered by CATV might expand to include data retrieval, video newspapers and magazines, banking, and home shopping.

In 1972, the FCC set in motion a gradual deregulation of the cable industry to permit the growth of its subscribers and programming. Cable systems had to carry all local market station signals. But, in addition:

- Operators in the Top 50 markets could carry up to three network and three independent stations.
- In markets 51-100, the rule was three network and two independent signals.
- In smaller markets, systems could bring in three network signals and one independent signal.

The new regulations also included what was known as the "leap frogging" rule. Very simply, it said that the stations brought into a market had to come from one or both of the Top 25 markets nearest to the cable system. An operator could not "leapfrog" over one market to pick a preferred station from another market. If this had been possible, a "Superstation" could have been born several years ahead of Ted Turner's WTBS in Atlanta.

While CATV operators had the green light to carry local signals and import distant ones, existing local market television stations (within 35 miles of the CATV operation) would receive considerable protection in the form of program exclusivity. For example, a CATV system could not duplicate local market programming via an imported signal from a distant station. It would have to "black out" such programming if it was also carried by a local station in the market. In terms of sports, the FCC also ruled that cable systems could not carry games that were blacked out on local network stations normally carried on the system. And, finally, cable systems in the Top 100 markets would be required to spend money to have a minimum capacity of 20 channels, to have built-in two-way capacity, and to provide the public and local government with at least one channel gratis for educational and communications purposes.

Even with this so-called "lifting of the freeze," the implementation of cable's expansion was still impeded somewhat by two factors. First was the considerable backlog of applications before the FCC. Second was the copyright controversy, for while cable operators had agreed in principle to copyright payments for the programs they carried, they and the copyright owners had yet to agree upon a specific system of fees.

Advertising Moves Several (Small) Steps Ahead

In this climate of potential future growth mixed with still-unresolved issues, 5,000 members of the cable industry met in Chicago in May 1972 for the 21st Annual Convention of the National Cable Television Association. There were now six million subscribers, and in an atmosphere of "happy confusion," the industry began to tackle the many problems that would need to be solved if cable was to become a significant communications medium and not just a small-town mover of signals. One such topic was advertising.

There was little to indicate that cable operators were really interested in the national advertising dollar in 1972. This was most evident in the absence of any significant number of program suppliers. The emphasis was largely on barter, movies, free films, or low-budget talk and interview projects. The most ambitious program offering was a Julia Meade series that offered advertising and would be available on cable systems with three million subscribers. (Remember that this was several years before satellite delivery of cable programming.)

There was discussion of CATV's potential on a PI (per inquiry) basis as electronic direct mail. For the most part, however, the real future of the medium as seen in 1972 was felt to lie in the area of strong local-interest shows. Advertising opportunities could develop here, and advertising agencies could take advantage by using their local offices, field representatives, and regional broadcast buyers as information sources.

Considerable attention was focused on pay television, with two systems announcing its introduction in San Diego, Sarasota, and Vancouver. The FCC wanted to encourage pay TV, but the real key to its success was a supply of product—namely big, new movies. Even at this still relatively early point in the expansion of cable, there was discussion about the limited use of commercials on pay television to divide shows into three acts and partially subsidize its costs, reducing prices to subscribers.

The relative lack of interest that cable operators displayed in advertising during the early 1970s was really not surprising. After all, CATV, unlike broadcast television, had a revenue source all its own—subscriber revenue. More important, the system operators were mechanically and not marketing oriented. They could string cable, but very few knew how to sell commercials, much less produce them.

Before cable could successfully attract and hold advertising revenue, it would have to document its ability to compete with other television opportunities in the efficient delivery of an advertiser's prime prospects. And cable operators would have to be shown how advertising could become a significant source of revenue and how they could develop it without either jeopardizing their subscriber revenue or putting in an undue amount of effort.

To put cable advertising at this time into perspective, in 1970 only 500 out of 2,500 systems were capable of originating programs; another 1,500 could provide automated originations such as time, weather, news, and stock ticker reports. Advertising was carried on only 57 of the cable systems that originated programming and on approximately 400 more that provided automated services. The average commercial cost about $15 a minute, and advertisers paid under $100 for an hour show. The best estimate of total cable ad revenue for 1970 was $3 million, a figure that compared with $300 million in subscriber revenue and well over $3 billion spent for broadcast television advertising.

Advertisers' use of cable followed a variety of paths:

- General Foods, Campbell Soup, Hudson Vitamin, and RCA Records tested cable in five small markets via "Monitel."

Owned 49 percent by Readers' Digest, Monitel offered a 24-hour feed of time, temperature, weather, general information, and national advertising. The test was inconclusive, and Monitel ceased operating.

- In New York, sports events, including the New York Knicks and Rangers games, were cablecast from Madison Square Garden on Sterling Manhattan Cable and Teleprompter. Avis, Schaefer Brewing, Miles, L&M Cigarettes, and Warner-Lambert each spent about $100 a spot for the coverage.
- In Pittsfield, Massachusetts, McDonald's sponsored the cable system's coverage of the annual Fourth of July Parade.
- On Lower Bucks Cablevision in Levittown, Pennsylvania, ads appeared 72 times each day at 30-minute intervals on the automated "TeleShop'r" service. The cost of spots in this feed of horoscopes, menus, and related incidental information was about 6 cents each.
- An early "how-to" show was a series of 13 instructional programs—"Salt Water Game Fishing." Teleprompter scheduled it three times a week on its Manhattan system and sold advertising time.

TABLE 1-2 Advertiser Interest in Cable, 1970

	National Advertiser	Local Advertiser
Automated time, weather, news	No	Yes
Older syndicated shows already run locally	No	Yes
Cable shows aimed at a small, select audience (How-To)	Yes	Yes
Local community programming	Some	Yes
Programs produced by larger MSO's (a while off)	Yes	Yes

These represented a few of the early ways in which advertisers were using cable. Table 1-2 indicates advertiser interest.

For the present, however, viable opportunities that would attract the national or regional marketer were extremely limited, and cable did not yet offer potential advertisers an efficient means of advertising. Furthermore, even when advertisers did purchase time in CATV programming to take advantage of the medium's selectivity, audience delivery was very difficult to evaluate.

Viewed from the early 1970s, the potential of cable was largely in the future. But, as a J. Walter Thompson Company report noted in January 1973:

> It is not too soon to be thinking of future uses which might enable advertisers to capitalize on the unique opportunities which cable will offer as a direct response medium:
>
> - With two-way capabilities, immediate response to advertising is possible. The vehicle might be a "how-to" show tied to a particular product or service.
> - The cable family can be identified and reached by at least one monthly mailing from the operator. Advertising material could be included within the mailing.
> - Cable could serve as a classified medium.
> - Cable could provide a research tool for measuring advertising communications effectiveness.

Considerable discussion was taking place about two-way cable, the growth of cable networks, and the development of pay television via cable, and its audience selectivity was on the minds of many. In September 1970, Benton & Bowles' top media executive George Simko said:

> There is the opportunity to develop and place programming that is geared to very specific audiences defined by job function, lifestyle, etc. On this basis, it is conceivable that television advertisers could isolate highly selective marketing targets in much the same way as selective magazines now perform this function. However, here again the broader programming availability could also serve to fragment the mass television viewing audience still further, leading to greater out-of-pocket expense required to deliver current levels of audience....
>
> If they could produce highly specialized programming,

programming of interest to doctors, let's say, advertisers might be willing to pay more than they do for other media and measure the cost on another yardstick. (*Marketing Communications* [April 1971], p. 31.)

By the mid-1970s, approximately 3,500 cable systems were serving about ten million subscribers. Unfortunately, however, the tremendous costs of wiring the major markets and the fact that most people there got relatively good over-the-air reception was holding back the growth of the industry. In addition, system owners who hoped to reap large revenues with Pay Television couldn't sell Home Box Office to someone who hadn't signed up for the basic cable service. Many large cable companies suddenly faced the prospect of losing vast amounts of money.

The Breakthrough: Satellites, Pay Services, and the Superstation

A breakthrough was needed—something that would transform CATV and its improved signal carriage into *cable television* with a vast new menu of viewing opportunities. This breakthrough occurred in 1975-76.

The first significant event took place on December 12, 1975, when the Satcom I communications satellite was launched. This provided a highly cost-efficient means of distributing multiple program options to cable systems across the country. Programming could be beamed from the ground 22,300 miles up to the satellite and then down to receiving earth stations (or "dishes") at cable systems across the country.

By providing instant reach of nationwide cable audiences, the satellite made it financially attractive to offer programming that would attract subscribers. It was HBO that first capitalized on this when in late 1975 it took a gamble and began distributing the first pay television mix of movies, sports, and specials via satellite to cable systems nationwide.

In those areas not yet cabled, pay television services were developed with distribution over the air. In some markets this was known as Subscription Television (STV). A scrambled television signal was sent over the air and unscrambled by a decoder box on the viewer's television set. In other areas, Multipoint Distribution Systems (MDS's) transmitted a pay service via microwave for generally short distances to subscribers who picked them up with special antennas and converter boxes. The interesting point is that although the long-term potential of "nonbroadcast television" was felt to be in its delivery of a wide variety of very special services and programming, it was the movies, with their widespread mass appeal, that really got things moving.

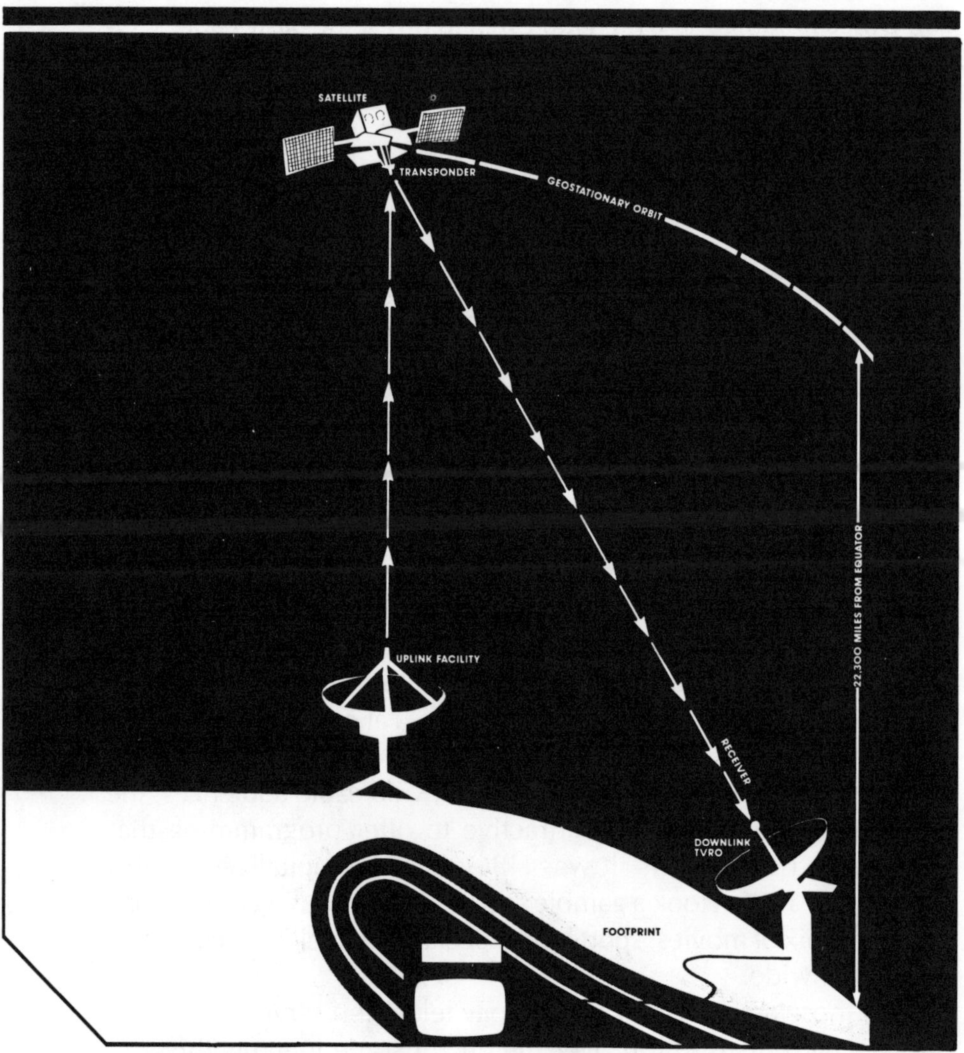

A basic satellite communications system

TABLE 1-3 Up on the Satellites

More than 40 video programming services as well as a growing number of stereo audio services are now available to cable systems. Viewers have a wide array of options from both the subscriber-supported and the advertiser-supported networks.

Premium Entertainment and Movies
American Movie Classics, Bravo, Cinemax, The Disney Channel, GalaVision, HBO, Home Theater Network, The Movie Channel, The Playboy Channel, Showtime

Superstations
WGN (Chicago), WOR (New York), WPIX (New York), WTBS (Atlanta)

Sports
ESPN, USA Sportstime

News, Business, and Weather
AP NewsCable, Business Times, Cable News Network, CNN Headline News, C-SPAN, Dow Jones Cable News, Financial News Network, Reuters News View, The Weather Channel

Ethnic
Black Entertainment Television, SIN Television Network

Entertainment and Performing Arts
Arts and Entertainment

General Entertainment, Information, Service and Lifestyle
CBN Cable Network, MSN The Information Channel, The Learning Channel, Lifetime, The Nashville Network, Satellite Program Network, The Silent Network, USA Cable Network

Religion
American Christian Television System, Eternal Word Television Network, National Christian Network, National Jewish Television, PTL Satellite Network, Trinity Broadcasting Network

Music
Country Music Television, MTV: Music Television, VH-1

Children's
Nickelodeon, USA Kidstime

Shopping and Participation
The CABLESHOP

While all of this was going on, Ted Turner, an imaginative, fast-talking businessman who owned the Atlanta Braves, and a 24-hour independent Atlanta television station, came up with the idea of using a satellite to distribute his station's programming to cable systems across the country. The only thing standing in his way was the

FCC's leapfrogging rules. This would prevent a cable system from taking his station if there was an independent in a Top 25 market nearer to him that could be carried. But in January 1976, the FCC eliminated the leapfrogging rules, and in December of that year, the nation's first Superstation, WTCG (now WTBS), was up on the bird and flying!

Thus, in only about a year, more steps had been taken to provide cable with the resources and impetus to leap ahead than had occurred in the first 25 years of its existence.

During the last third of the 1970s, earth receiving stations rapidly popped up at cable systems across the country. Their growth was spurred on by sharp reductions in their cost and by a rapid acceleration in the number and variety of satellite services. This, in turn, sped the household growth of existing cable systems and increased the pace of new cable construction throughout the country. (See Table 1-3.)

Cable was ready to enter the 1980s!

Charles Barsotti, © 1984

CHAPTER 2

The New Media: Today and Tomorrow

Television of the 1980s is a far cry from that of earlier days. Sets are available that fill a wall or fit into a pocket. They can be plugged into a car's cigarette lighter or operated by a battery. They even have telephones built in so you can watch and talk at the same time. In the 1980s, what has been known simply as "television" is becoming "viewer-controlled video." In the process, viewers find themselves no longer prisoners of what is on and when it is on. Rather, they watch what they want when they want it with dozens of program choices and a variety of video recording and playback devices.

Let's briefly examine what elements make up the new video mix.

Cable and Pay Cable

Cable television today reaches over 40 percent of the homes in the country. Of all the new media forms, it is today the most dominant in terms of viewer penetration and advertising potential. Until recently, cable was largely a small-town, at most a 12-channel, proposition used to bring TV to rural areas and improve reception. This is rapidly changing as the cable companies franchise the big U.S. cities and construct 36-, 52-, and even a few 104-channel systems.

Cable subscribers pay a monthly fee of about $8.00 to supplement their broadcast television diet with a wide variety of special programs. For approximately an additional $9.00 a month, subscribers may purchase what is referred to as a "pay cable" service. This consists mainly of movies with some sports and special programs—uncut, uncensored, and (at least for now) uninterrupted by commercials. Many cable systems offer "tiers" or combinations of services packaged at different prices. Homes subscribing to them in essence become "multiscreen cinemas."

Subscribers to pay cable services have long been regarded as representing very lucrative marketing targets. They are younger, better educated, and substantially more affluent than the population at large. As a result, considerable interest has been expressed by advertisers and advertising agencies in having commercial opportunities available on pay cable.

It is debatable whether or not the major *existing* pay services will become advertising carriers, even if the commercials are uninterruptive (between rather than within films) and of a unique, entertaining nature. The pay services realize that one of their attractions to many subscribers is their noncommercial nature. If only 1 percent or 2 percent of their subscribers cancelled because of the commercial intrusion on them, the services and their systems could lose more in revenue than they might ever expect to recoup in advertising income. (See Table 2-1.)

New pay cable services, however, may well begin operations with a limited amount of advertising and offer subscribers a lower monthly cost than *existing* services. There also are advertising opportunities in the natural breaks in pay sports. In addition, advertisers will probably find numerous opportunities to reach pay subscriber homes by advertising in pay cable program guides. It will be a case of using a print medium to reach a video audience.

STV, MDS, and SMATV

In large cities where cable television has not yet arrived, pay services are often available via Subscription Television (STV) and Multipoint

Distribution Systems (MDS). These are single-channel services for which subscribers pay about $20 per month plus an installation fee.

Signals are delivered over the air via one of two techniques:

- Subscription Television involves delivering a scrambled picture from a UHF station that is unscrambled in subscribers' homes via a decoding box. This is the technique that ON TV offers over Channel 44 in Chicago.
- A Multipoint Distribution System also employs an over-the-air system, but rather than coming from a UHF station, it comes from a closed-circuit microwave transmitter. Again, a scrambled signal is unscrambled by the decoding box on top of the set, but this time it is played through an empty channel. In Chicago, Showtime is an MDS system available to dwellings within 10 miles of the Hancock building.

STV and MDS systems are available today in only 1 percent of the nation's households. Both industries have suffered in recent years with many systems shutting down completely and others losing large numbers of subscribers. The industries' problems can largely be at-

TABLE 2-1 The Growth of Cable

	Basic Cable			Pay Cable	
Year	Cable Systems	Subscribers (millions)	Percent of TV Homes	Subscribers (millions)	Percent of TV Homes
1952	70	.014	>1	—	—
1960	640	.550	1	—	—
1965	1,325	1.275	2	—	—
1970	2,490	4.500	8	—	—
1975	3,506	9.800	14	—	—
1976	3,681	12.100	17	.600	1
1977	3,832	13.200	18	1.500	2
1978	3,875	14.200	19	3.000	4
1979	4,150	16.000	21	5.300	7
1980	4,225	18.700	24	7.800	10
1981	4,375	22.600	28	11.800	14
1982	4,825	29.300	35	17.000	20
1983	5,600	34.100	41	20.800	25
1984	6,000	39.000	45	24.000	28
1990	7,000	56.000	58	38.000	39

Source: Extrapolated from *Television Digest, Cablevision, Media Science Reports* tributed to the growth of cable, whose subscriber base more than doubled between 1980 and 1984. In contrast, according to industry analyst Paul Kagan, the STV and MDS subscriber base dropped from 2.1 million in 1981 to 900,000 in 1984.

Unlike cable, which brings a large multichannel menu of programming into a home, STV and MDS systems represent a single channel with movies and sports as the major attraction. For the most part, STV and MDS homes can be expected to convert to cable as soon as it is available to them.

In a number of major markets where cable may not yet be available, still another service has been developed—Satellite-Fed Master Antenna TV. SMATV is a minicable system for buildings connected to a private satellite antenna. It provides multichannel video to large apartment buildings and condominium complexes and may "syphon" some potential subscribers away from cable in markets that are slow to wire-up.

Pay-Per-View

There are strong indications that an area of high potential in terms of future movie and special event exposure is what is referred to as "pay-per-view." Since homes must be specifically "addressed" and fed individual programs for which they are charged, it requires special equipment.

However, it is quite possible that television viewers soon may be able to pay to see one-time showings of films that have just completed their initial runs in theaters. The distribution pattern that develops might be: movies would be released to theaters first, then broadcast on a pay-per-view basis, followed by home distribution in video cassette and disc form. Next would come pay cable and STV, SMATV, DBS, and MDS. Only after these stages would some of the films come to free broadcast television (obviously with significantly reduced audience potential). (See Figure 2-1.)

Previous exposure of major theatrical motion pictures on pay channels as well as on video cassettes and discs is already cutting sharply into their ratings when they premiere on network television.

The network premiere of "Star Wars" drew only a 35 percent share of audience when it aired in February 1984. And "Superman II" earned only a 27 percent share for its first network television run

FIGURE 2-1 Possible Entertainment Distribution Pattern of the Future

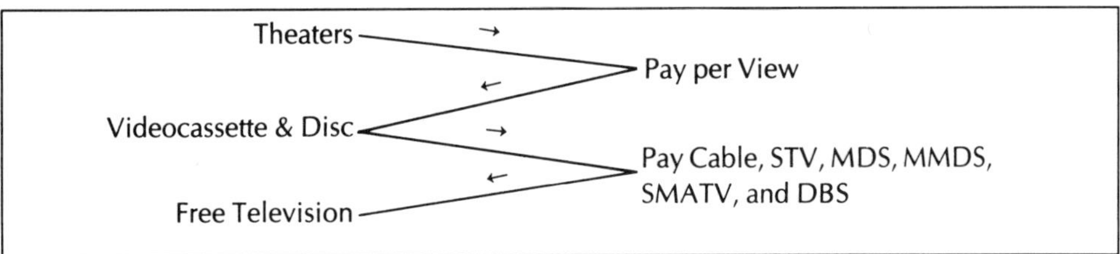

that same month. While both of these films would have easily delivered 50 to 60 shares in the "Pre-New Media Days," the influence of pay, tape, and disc exposure prior to their network exposure was enormous.

There also is the potential of high school sports to deliver a large pay-per-view audience. On November 5, 1983, in Cincinnati, Moeller High School and Princeton High School were playing for the city championship, and the football stadium had sold out all of its 10,000 seats. Warner Amex cable got permission from the schools and state officials to offer the game on a pay-per-view basis and reached an audience of more than 6,000 at $5.00 per home. The result was a profit and a potential for televising high school football for money in future systems. The success of pay-per-view high school sports is bound to attract the attention of local auto dealers, food franchisees, and other retailers as they consider the advertising and promotional potential of these events.

Multichannel MDS

Multichannel MDS services would challenge cable by providing more than a single entertainment channel. As "an urban over-the-air wireless cable network," they could offer two-way data services plus pay-per-view movies, in addition to the usual entertainment, information and sports. Fees charged subscribers would be about the same as cable with a one-time charge for a special antenna.

A multichannel MDS system would provide a new service in noncabled areas *and* competition to cable operators in wired areas. It would also be an additional "viewer competitive force" with which local broadcast stations would have to contend.

Satellites

On September 30, 1975, Home Box Office transmitted to its subscribers the Ali-Frazier heavyweight fight live from the Philippines. The unusual thing about this transmission was that it was via Western Union Westar satellite (or "bird") locked in geostationery orbit, 22,300 miles over the equator. The impact of communications satellites on the cable industry has been dramatic. In 1984, some 50 program services are being transmitted to receiving earth stations at cable systems across the country by RCA's Satcom, Western Union's Westar, AT&T's Comstar, and Hughes Galaxy satellites.

Satellite transmission and its diversity of program offerings have had a major impact in spurring ahead cable franchising in the major markets. And, as the cost of earth stations has come down, TVRO, or television-receive-only satellite dishes are popping up all over the country. With the price of a dish dropping to under $2,500, estimates placed the number of home earth stations at 800,000 in early 1985. It's the latest status symbol!

Direct Broadcast Satellites

Direct Broadcast Satellites (DBS) send a signal from a satellite directly into individual homes via small two-or-three-foot rooftop earth stations. With a decoder, an individual can pick up four or five channels that feature movies, sports, the performing arts, children's shows, and other special interest programming. DBS offers more variety than single-channel STV and MDS but substantially fewer viewing options than cable. Unlike cable however, the enormous investment of stringing cable will be avoided. Still, the cost of setting up a DBS system is not inexpensive. It is estimated that it will take as much as $1 billion for a national system!

The DBS era officially got underway on November 15, 1983, when United Satellite Communications, Inc. commenced a test serving an area in central Indiana. Over the following months, the service expanded into several thousand homes in the Midwest and Northeast. The costs involved in starting up a DBS service coupled with uncertainties as to its future have resulted in major cutbacks in the number of companies planning on getting into the business. CBS, Western Union, and RCA all pulled out of or postponed their ventures.

Many questions remain as to whether the quality and variety of programming together with the pricing of home receivers (about $300) will make DBS a true competitor of either broadcast or cable television.

From a practical standpoint, the future of DBS is largely dependent on the number of customers who would pay $25 or so a month for the service among the estimated 15 million rural and noncable homes that would be its prime market. DBS will initially focus on these areas. In the long run, however, it also may offer considerable potential to entertain people in "nonresidential" areas, such as restaurants, in waiting rooms in hospitals, at clinics, in auto repair shops, etc.

Low-Power Television

While the stakes involved in getting into cable or DBS are enormous, a plan by the FCC may enable an individual to establish a television station for well under $100,000. The technology behind what is known as low-power television (LPTV) is not new. For many years, low-power transmitters have amplified and rebroadcast the weak signals of distant major-market stations to viewers miles away in rural areas. Now, the FCC is permitting the use of this technology to allow the building of low-power, or mini, television stations with broadcast ranges limited to 10 to 20 miles. The FCC hopes this will open up the commercial airwaves to women, minorities, and other new voices.

From an advertising standpoint, low-power television could offer opportunities for targeting very specific messages to some very specific audiences. In a sense, it could develop much like local radio with thousands of stations nationwide. However, only a few LPTV stations are operating today, most in rural areas, and low-power television is growing *very slowly*.

While the pinpointed geography of LPTV might seem appealing to advertisers, the sheer cost and complexity of negotiating with thousands of small stations should rule it out as a viable national medium. Low-power television operators will find it far more profitable to concentrate on local and regional accounts.

In the long run, low-power television networks designed to serve predominately rural or suburban areas might offer national potential. For the short term, however, any national business that is developed will probably focus on direct-marketers seeking immediate response to their 800 numbers.

Video Cassettes and Video Discs

Up to this point this discussion has covered only new media programming that is transmitted, in one way or another, to a viewer's television. Video cassettes and video discs are two forms of new media that do not involve direct transmission to an individual's television.

The video cassette recorder (VCR) plays prerecorded material (largely movies that can be bought or rented) and, when used with a tape camera, can take electronic home movies. The newest 8mm video cassette format uses approximately a quarter-inch tape, versus the conventional half-inch. It is fully portable and combines a camera and cassette recorder in one lightweight unit.

Unlike the VCR, current video disc players cannot record. Like a phonograph, which the discs resemble, they can only play back. The more expensive laser models have stereo and special features such as freeze-frame, frame-by-frame advance, slow motion, fast motion, and fast scanning. The less expensive stylus models do not have all of these features. And, of course, as is the case with video cassette recorders, the recordings on one system cannot be played on another "incompatible" system.

Since the video disc player cannot record, what would be its advantage over the VCR? *First,* it is the cost of the discs relative to tapes. A movie might cost twice as much on tape as on disc. Since most people rent tapes at $1 to $5 a day, this may become academic. The *second* possible advantage is the outstanding picture quality and, in the laser models, the stereo sound.

On the optical (laser) disc, there are 54,000 still frames that can be stored on *each side.* With the ability to access each of these frames individually, there is an enormous number of ways in which the disc might be used in the dissemination of information. The entire works of the great artists of all time, encyclopedias, dictionaries, travel

VHS
THE PHENOMENAL NATIONWIDE BESTSELLER!
#1
EAT TO WIN
THE SPORTS NUTRITION BIBLE
STARRING
DR. ROBERT HAAS
AUDREY LANDERS
IVAN LENDL
KARL HOME VIDEO
EAT TO WIN
DR. ROBERT HAAS

guides—all these and much more can be stored on very few discs. Their marketing potential was recognized by Sears when it developed a video disc version of its Summer 1981 Catalogue that allowed customers to study stills of the products as well as moving-action sequences and, of course, all of the necessary prices and specifications.

During 1983 and 1984 the home video industry boomed with one out of five households now owning VCRs. The prices of prerecorded tapes dropped sharply to $39.95 for many new releases, and some fell below $25. These trends all point to a very bullish outlook for home video as a new media form.

As home video grows, we may well expect it to attract the interest of advertisers. The movie studios have already attached trailors of upcoming films to tapes being sent down the line just as they do for theatrical release. And the record industry may see this as a new way to sample current releases through video clips. Advertisers may create "short subjects" to run at the start of a movie tape. Kodak, for example, could do one on "Great Photography Around the World," or Miller Brewing Co. could produce a short on "Great Sports Heros." Some original home video programming might even be 100% advertiser underwritten.

A media/advertising breakthrough in the home video market came in early 1985 with Red Lobster Inns' sponsorship of the best-selling book, "Eat to Win." Red Lobster Inns' identification was on the cassette package, on the actual cassette and on billboards preceding the program. No commercials interrupted the body of "Eat to Win" itself.

The Videophile is a movie preview program distributed free of charge to video stores nationally. It is designed to interest customers in the newest home video releases and sells 30-second spots to advertisers seeking to reach the upscale home video owners who tend to erase commercials from programs they have recorded.

Video Games

Today's video games have come a long way since the introduction, a few years ago, of Odyssey's Pong games with the bouncing dots. Programmable units, now in more than 1 out of 5 homes, allow a person to choose from dozens of games with incredibly realistic representations of space ships, playing fields, and games and players, as well as great control over strategy. It is not even all fun and games any more, as manufacturers are now offering educational cartridges.

A major development in arcade games were the laser-disc games, with "Dragon's Lair" starting the ball rolling in the summer of 1983. Combining traditional video game technology with animated cartoon or live action sequences, these games enable players to actually control the film sequences and hence the action they see.

Bally Manufacturing Corp. joined with RCA Corp. in early 1984 to introduce "Bally's NFL Football," a laser-disc game for taverns as well as arcades. The RCA video disc stores films of actual professional football plays. The players representing either the offense or defense, choose strategies which determine which plays are flashed on the screen. If successful, this could introduce video games to an older age group than those that play in arcades.

It would not be at all unusual to see advertising built into such a game in the form of billboards visible on the playing field, or on the "scoreboard" or even an actual mini (three-second) commercials at the start and end of the game and even at "half-time." If they were entertaining and if they helped to hold down the cost of the games, they might well be accepted by the players. And, of course, the advertiser might even underwrite a competition, such as an auto manufacturer giving a new car to the top scorer.

Bally and Anheuser-Busch's Budweiser teamed up to introduce "Tapper," an adult video game. The action of the game is a carryover from one of the oldest barroom amusements around—where the bartender slides a mug of beer down the bar without spilling a drop. The same skill must be executed in "Tapper," with the player at the controls sliding the mugs electronically toward a target. And, of course, Budweiser is the beer in the mug.

A new merchandising tool applies the latest in video game technology to point-of-purchase displays with the result being an interactive in-store video system. VideoSpond combines an optical laser videodisc (as in the "Dragon's Lair" arcade game) with a microcomputer and printer in a self-contained information center. Consumers select products about which they desire information and determine just how much information they want to receive. They also receive an audiovisual product demonstration, and a built-in printer can produce product information material, such as recipes or coupons.

By providing more than is available in traditional 30- or 60-second commercials, VideoSpond brings in-depth information to the point-of-sale for marketers in such categories as hardware, automotive, fashion, travel, and food products.

Personal Computers

Personal computers were developed largely for small businesses headquartered at home and professionals such as doctors, lawyers, and dentists. By early 1985, however, more than 1 out of 10 households were using home computers to organize financial records, do taxes, store information, play games and as word processors. While little has been done to develop advertising on home computers the potential undoubtedly is there. In particular, "how-to" programs with product tie-ins might be offered on an almost limitless variety of subjects.

H&R Block could offer a program for maintaining home financial records. At the end of the year, it would be delivered to the nearest H&R Block office to be used by H&R Block in preparing an individual or family's tax return. No more last-minute gathering together of records on scraps of paper!

Predictions for the increased growth of the personal computer business are very bright. In his Fall 1983 *New Electronic Media Projections Report,* Bill Harvey, publisher of *Media Science Reports,* forecast that by the end of 1990, 4 out of 10 U.S. homes would be using them—with 24 percent owning and 16 percent renting.

Interactivity: Qube and Videotex/Teletext

The last half of the 1980s will be known as the Era of Interactivity. As new wiring is laid and smaller 12-channel cable systems are upgraded to 36, 52, and a few 104 channels, two-way activity will expand.

The first interactive cable service was Qube. Qube was developed in Columbus, Ohio, by Warner Cable. A marriage between the television set and the computer, Qube enables viewers to respond to messages on the screen. They can answer questions asked on the screen, take part in instant polls, and even purchase products from their living rooms.

Another category of interactive information transmission services actually includes two services—teletext and videotex.

In teletext, alphanumerics and graphics are carried over the unused picture linage of a television channel and delivered page-by-page as selected by the viewer. On a regular TV channel (such as KNXT in Los Angeles, one market where it was tested), teletext has a capacity of about 200 pages. When it is given its own full cable channel, the capacity grows to about 5,000 pages.

On the other hand, with videotex, the alphanumerics and the graphics are stored in a computer and delivered over telephone lines or cable. The viewer can select this material and interact with it. The potential number of pages of data that can be generated is limited only by the capacity of the computers employed.

Viewtron completed a test of a videotex system in 1981 in Coral Gables, Florida, and began expanding it to 5,000 south Florida homes in November 1983. The tests are underwritten by Knight-Ridder with technology supplied by AT&T. The system is simple to operate and allows subscribers to select what they want through the conventional "branching-tree" process, a constant narrowing down from broad categories of information to more and more precise data. The user has a key pad that is used like a telephone to communicate with the system and request information. Or, requests can be made by using a typewriter-like keyboard to request, say, W-E-A-T-H-E-R or D-O-W I-N-D-E-X.

Information transmission services have considerable advertising potential for many companies who want to transmit in-depth data about their products or services to consumers on an "on-demand" basis. Videotex can be used for direct home shopping in which the home customer can select what he or she wants to purchase from a variety of options and actually place an order.

Bigger Screens, Better Pictures, Sensational Sound

Just as changes are taking place in what can be done on the television screen, so are there significant innovations in the hardware itself.

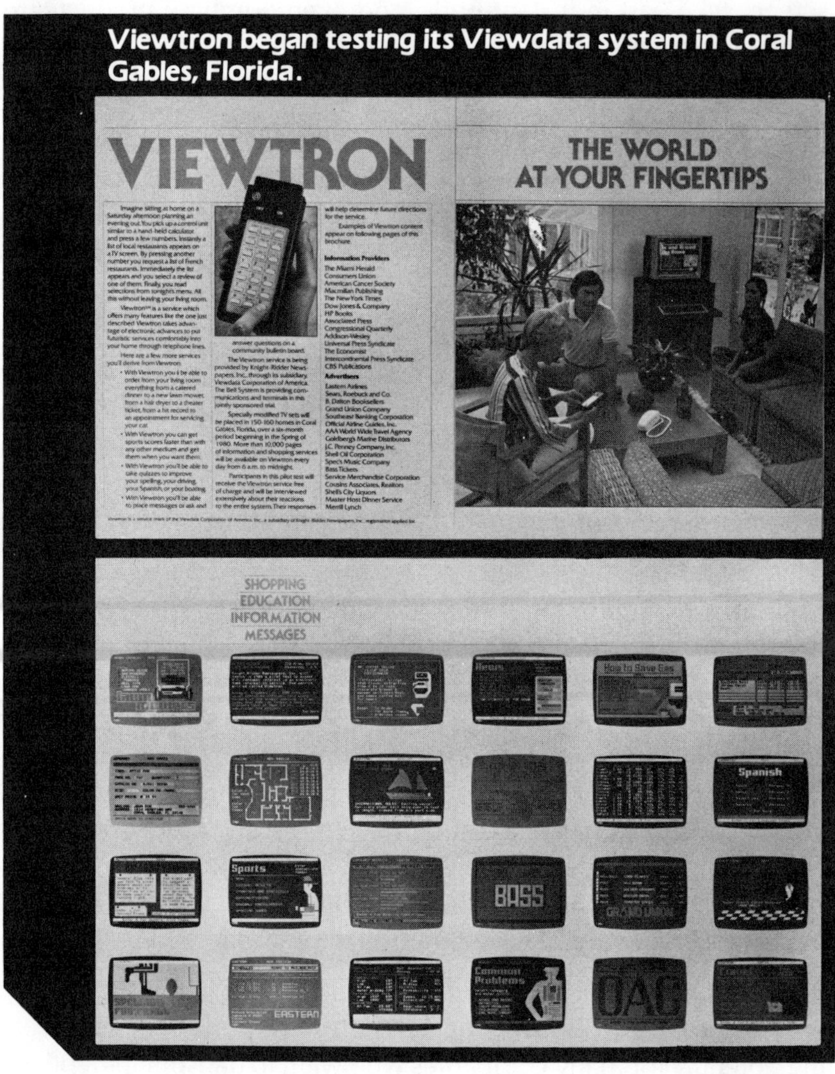

Viewtron began testing its Viewdata system in Coral Gables, Florida.

Matsushita Electric, Japan's largest consumer electronics company, has been developing:

- A TV screen 8½ x 11 feet that can be used for either front or rear projection.
- High Definition Satellite TV system for receiving super-high-frequency (SHF) signals transmitted simultaneously to the entire country. With 1,125 scanning lines rather than the 525 currently in use in the U.S., the picture quality is far sharper than what we now see.
- Television with stereo and bilingual sound.
- Three-dimension video cassettes in which the viewer wears stereoscopic glasses to see the effect. An airline might use this to develop an entire series of 3-D travel films for promotion purposes.

FIGURE 2-2 The Branching Tree Process

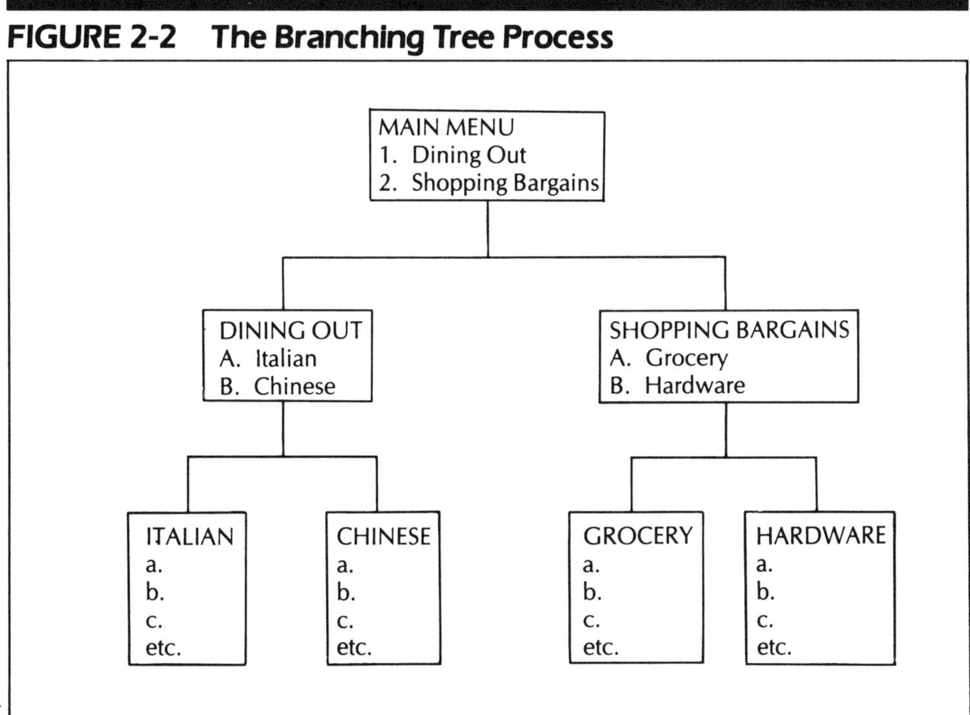

A Multitude of Choices

The 1980s will be known as the "decade of media choice." New video hardware and a variety of video software will combine to expand both the number of options available to the viewer and the flexibility with which he or she can exercise these options.

In his detailed analysis of new media—"Projections of NEM Rollout through 1995"—Bill Harvey, publisher of *Media Science Reports,* commented:

> Eventually, most Americans will feel incomplete without practically all of the new electronic media. These 21 NEM "products" and the innumerable incremental services they support are the main growth engine to advertising for at least the next 30 years. (See Table 2-2.)

TABLE 2-2 A Roster of New Electronic Media

Cable	Videodisc Player (VDP)
Personal Computer (PC)	Direct Broadcast Satellite (DBS)
Pay Cable	Home TV Camera
Remote Control	Highspeed Receive-Only PC/VG Downloader
Videogame (VG)	Multipoint Distribution Service (MDS)
PC-Based Viewdata	TV Downstream/Telco-Upstream Viewdata
Videocassette Recorder (VCR)	Picturephone
Two-Way TV	Camera Two-Way TV
Teletext	Subscription TV (STV)
Two-Way-Cable-Based-Viewdata	Low Power TV (LPTV)
Giant Screen	

Source: Bill Harvey, *Media Science Reports*

At the end of 1983, Harvey had estimated that half of U.S. television households already had one or more of these 21 types of hardware in their home, with the average home having 2.5 NEM types. By the end of 1990, Harvey predicted penetration at nearly 8 out of 10 homes, with an average of 4.0 NEM types per home. And by the close of 1995, he estimated the numbers would reach near saturation (98.0 percent) with the average home having 1 out of 4 (5.3) new electronic media devices.

In a nutshell, by the time the decade of the 1980s has ended, we will be looking at the widespread blooming of many new media that have only just begun to sprout.

CHAPTER 3

The 1980s and Beyond: New Options for All

Advertisers today must become knowledgeable about the new electronic media from the standpoints of diversity of viewing choice (new options for using the TV set), diversity of viewing time (time-shifting potential), and interactivity.

More than anything else, the 1980s will offer advertising prospects a new *diversity* of video experiences. With diversity of *choice* of both information and entertainment, consumers will no longer be restricted to a handful of offerings available from the networks and independent stations. And diversity of *time* means they will no longer need to alter their lifestyles to fit network or station schedules. Viewers will watch what they want to see when they want to see it *or* are available to see it.

Diversity of Choice

Viewer choice has steadily increased over the past 30 years as a result of the growth in the number of broadcast stations and cable systems. In 1983, the average home could receive more than two and one-half times the number of over-air TV stations it did in 1953. (See Table 3-1.) Looked at another way, over half the homes in the country could

TABLE 3-1 Over-Air TV Stations per Home per Year

Year	Average Number
1953	3.8
1960	5.7
1970	6.8
1983	10.3

Source: A.C. Nielsen, 1983

receive 10 or more stations in 1983, while 29% received 13 or more. (See Table 3-2.)

And this is just the beginning of choice, since it reflects only over-air TV *stations*. According to a September 1983 report by the ICR, Cable Information Service (Titsch Communications Division), 60 percent of cable subscribers could receive 30 or more *channels*, and 80 percent could receive 20 or more. New cable systems now under construction offer more than 50 channels. And the older 12-channel systems will eventually be upgraded to provide the same larger variety of selections. Hence, it is not out of line to forecast that, by the end of the decade, viewers will be able to choose from a vast array of video offerings in the same manner as they now select what to listen to from a multitude of AM and FM radio stations.

Of course, for those desiring even greater choice, video cassettes and discs offer additional alternatives just as stereo records and tapes provide audio diversification.

TABLE 3-2 Number of Stations Received

Number of Stations	Homes Receiving (Percent)	
1-4	5	
5	5	
6	9	
7	9	
8	11	
9	10	
10-12	22	} 51
13+	29	
	100	

Source: A.C. Nielsen, 1983

"You want to know how much TV I watch each day? On which channel?"

FIGURE 3-1 Channel Availability in Cable Households

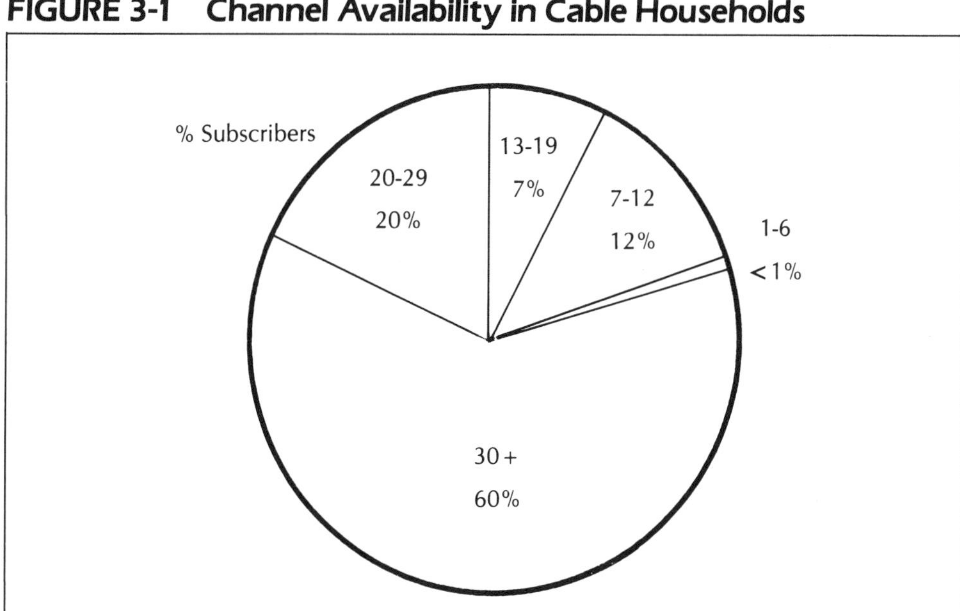

Diversity of Time

In Quincy, Massachusetts, a local ordinance forbade dancing on Sunday. Hence, the town's "California Disco" owner decided to show five straight hours of "General Hospital," which he had taped earlier in the week. Working women—whose jobs kept them away from their favorite soap operas Monday through Friday—packed the place. They spent $10 to $12 each on drinks, about the same as did the Saturday night disco crowd. Business was so good that the "California" expanded to two showings a Sunday!

This 1980 "happening" is a classic example of new media "time shifting." And among the growing number of homes that have video cassette recorders, viewing patterns are greatly altered.

The VCR gives its owner a "time shift" capacity by which he or she can record a program at one time and play it back at another. Days and hours blur together as viewers watch at their leisure and on their own time schedules rather than on the schedules of the media. And cable satellite networks include multiple exposure patterns—with programs and features repeated within the same day, week, month, or over the course of a year.

In homes with video recorders, viewing patterns are both altered and shifted as owners buy prerecorded programming and tape other programs to watch at times other than those of the original

FIGURE 3-2 New Options for Television

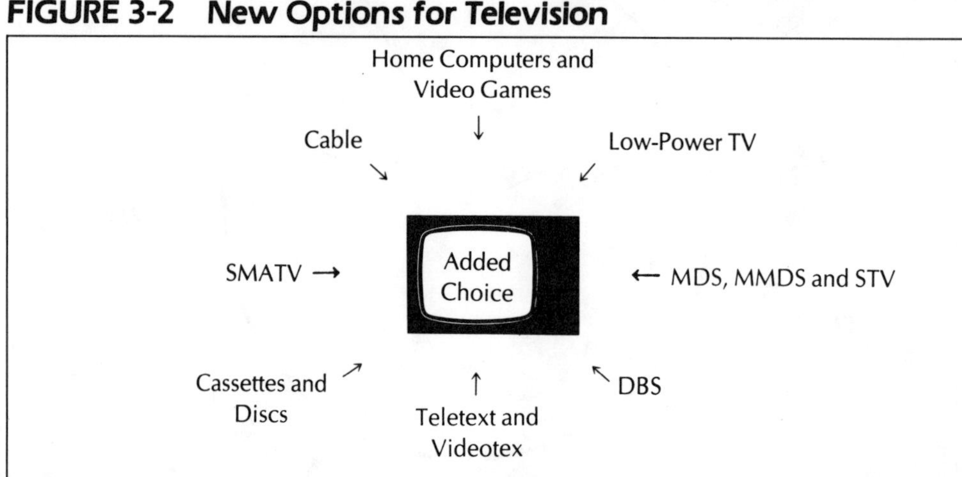

broadcasts. One important point is that, as the Nielsen Home Video Index reported, VCR households are upscale in education, occupation, and income. Equally important for advertisers is that Nielsen and other researchers have found that people would accept advertising in video cassettes if the prices of the cassettes could be held down.

Table 3-3 gives an example of how a working woman might create her own video viewing schedule. Over breakfast, she watches Johnny Carson's monologue from "The Tonight Show" of the previous evening. After dinner, she scans her tape of "The CBS Evening News," watching only those items which interest her—fast-forwarding past everything else. Later, she settles in with Phil Donahue, after which she curls up in bed with "General Hospital." With her VCR she has totally time shifted the network and station schedules to meet that of her own lifestyle.

Television viewing may actually develop a bit like magazine reading. People usually do not sit down and read an issue cover to cover. They pick it up at first, look at whatever interests them, and then perhaps return to it later and look at other material. With more viewing alternatives, these same people will be able to look at what they want to see and record other programs to look at in whole or part later.

By the end of the 1980s the claim that television reaches people at one single point in time might no longer be the case for all shows. Just as a magazine accumulates its audience over several weeks (or months), viewers will record certain shows and their audience will accumulate over time rather than in a single instant.

With the growth in VCR ownership across the country (20 percent of all homes by the middle of 1985) advertisers have two reasons for concern. *First,* the deletion of commercials by VCR owners at the time they record programs or during playback reduces the value of advertising in a show. And *second,* when the VCR is used to time-shift, there is a serious threat to the viewing of targeted, time-sensitive campaigns.

How serious is the problem? A 1983 NPD Electronic Media Tracking Service study (*Advertising Age,* October 31, 1983) found that 68 percent of VCR owners said they deleted commercials at the time of recording, and three out of five skipped past the ad messages during playback. In addition, 93 percent of the recordings were time shifted so they could be played back at a later date. And a 1984 A.C.

Nielsen study (*Electronic Media,* July 26, 1984) reported that 64 percent of all VCR playbacks were fast-forwarded over commercials. Obviously, advertisers will want to track this phenomenon closely.

TABLE 3-3 A Possible New Media Viewing Schedule

	Conventional Television	Time Shifted Video
Morning	"Phil Donahue"	"Tonight Show"
Afternoon	"General Hospital"	—
Early Evening	"CBS Evening News"	—
Prime-time	"Whatever is On"	"CBS Evening News" "Phil Donahue"
Late Night	"Tonight Show"	"General Hospital"

Interacting with the TV Set

As viewers gain a new diversity of choice and of time, they will also change the manner in which they interact with the television set. The typically stereotyped *passive* viewer who only watched (sometimes inattentively) will become an atypically *active* consumer of a vast menu of video offerings.

Through a number of data transmission systems, information will be delivered to the viewer on request and responded to by him. Such capabilities are with us right now and many are being tested.

Home Shopping

Despite its seemingly utopian quality, home video shopping is on its way into our living rooms. Inflation, transportation costs, and crime are all increasing the potential of this new video service. In addition, more husband-wife working households, coupled with the rise of the single adult household, are increasing the need for greater convenience and preservation of one's leisure hours.

There are obviously some products that the consumer will always want to see in person and examine in detail, but there are others that he or she would just as soon buy from home. It is possible that home shopping could drastically alter the retailing structure for a

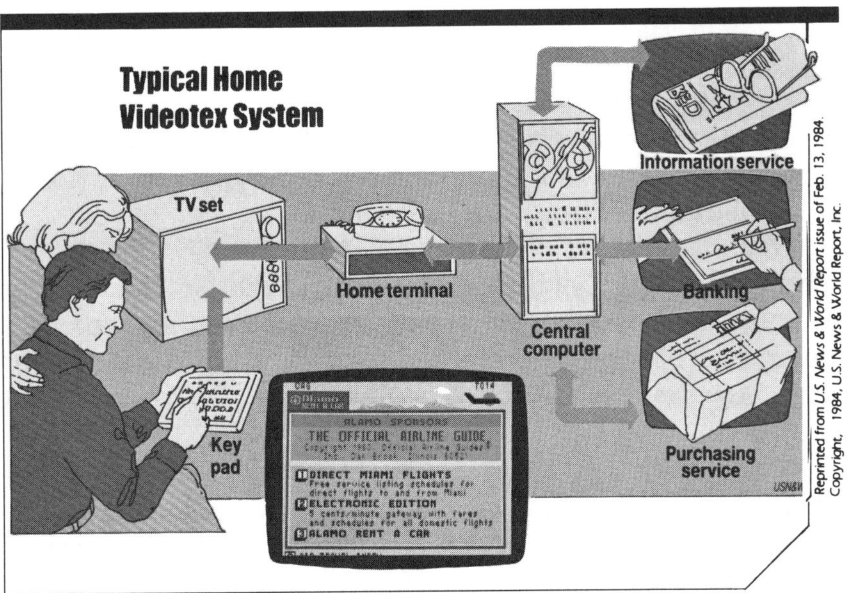

wide variety of these goods and services. For example, an aisle in the supermarket of the future might be a videotex video screen in the home. For perhaps the first time ever, a person can tell at a glance what everything costs. And, of course, unit pricing will also be included. In addition, on every frame will be the name of the retailer in whose "Video Supermart" one is shopping.

The "Video Shopper" was an example of a home shopping service tested a few years ago. It operated without two-way cable in a relatively simple manner. Consumers received the "Consumers Video Shopper Catalog." In addition to the information provided in the catalog, "Video Shopper" infomercials were shown on a local cable channel. These infomercials provided additional information on the product and explained special features. To make a purchase, the customer called a toll-free number and placed the order.

Home Security

An example of what some people feel may one day be a highly demanded cable service is home security. Using interactive equipment, it monitors subscriber homes for smoke, fire, or intrusion. There also is a "panic button" available in the case of a medical emergency. The entire system is linked to appropriate agencies and medical facilities.

Another special service is the Emergency Warning Device. When the community-wide alert system is activated, the device sounds in the home. In effect, it tells subscribers to turn on their television sets and tune to a special channel for emergency information.

While these services do not offer on-cable advertising potential, they do provide opportunities for tie-in print promotion. Companies such as Bristol-Myers, Abbott Laboratories, or Johnson & Johnson are ideal candidates for such advertising. Companies in these product categories could prepare home safety and first-aid pamphlets for local cable systems to include in their monthly statements to subscribers.

Interactivity Today

The impact of interactivity will be most felt during the last half of this decade when the hardware needed for it is more widely available to homes across the country. In the meantime, a number of interesting experiments are taking place to test its potential now.

One such project has been Adams-Russell's The CABLESHOP, with the telephone serving as the request device rather than a special cable converter. This interactive information service was first tested on an Adams-Russell 7,000 subscriber cable system in Peabody, Massachusetts from March 1982 to January 1983. Based upon the success of this test, it received a national workout on seven systems, with approximately 250,000, subscribers between November 1983 and April 1984.

The CABLESHOP programmed three- to seven-minute segments with helpful service-oriented, informative commercials, 24 hours a day. Viewers wishing to participate in The CABLESHOP had a "menu" on their screen and a program guide listing dozens of subjects from which to choose. They simply called up the particular item they wanted to see by dialing a special number and code on their telephones. A computer programmed their request so it came up on one of two CABLESHOP information channels within the following few minutes. Among the "infomercials" were what to look for in buying a car (sponsored by an automaker), new recipes (from a food manufacturer), and information on how to save and/or invest money (from a financial services company). As people used their sets to request information, they were no longer passive viewers but, rather, active video consumers.

In 1985, the CABLESHOP is scheduled to become a national satellite delivered "consumers' channel." The two-way interactive capability will be eliminated, since once the novelty wore off, few viewers used it. And in addition to the advertiser-supported infomercials, The CABLESHOP will carry its own consumer news and information.

For advertisers, The CABLESHOP offers the chance to test and measure results for selling a product or service directly, generate qualified sales leads, build in-store traffic, provide couponing and sampling, and provide support for print and direct mail advertising.

Perhaps the greatest areas of interactive opportunity involve the merging of a national marketer with a local retailer. Such a turn-key effort offers the opportunity to provide: in-depth product or service information with how and where to get more information or make a purchase.

Shopping at home with the CABLESHOP.

The CABLESHOP encourages such local involvement. In addition, a number of innovative local cable systems across the country have begun to develop imaginative interactive shopping and information services in their areas.

Starting now, Peabody Cable TV subscribers have something nobody else in the world has.

A Cable TV Shopping Information Service that helps you be a more effective shopper because you'll be a more informed shopper.

You'll see messages about products and services brought to you by some of America's leading companies. Recipes and household hints. Product demonstrations and do-it-yourself ideas. News of local sales at supermarket, drug and department stores.

And using The Cableshop is as easy as one, two, three:

1. Select a message. Every month your Cable Entertainment Guide contains a special section listing all The Cableshop messages. Select the message you'd like to see. Note the code number next to it. Since your message may have already been selected, turn to The Cableshop channel (51) to see if it's listed.

2. Phone in your selection. Dial The Cableshop number: 532-4372. A recorded message will ask for your Cableshop User number (sent to all subscribers through the mail), and the code number mentioned above.

3. Sit back, relax and watch. Your message is now scheduled. Watch The Cableshop channel (51) for the exact time and channel it will be shown on.

Here's what you'll see on The Cableshop channel (51):

Time of Day
Cableshop Phone Number
Message Code Number
Message Title
Channel that Message will Appear On
Time it will Appear

The Cableshop is available 24 hours a day. And it's free to all Adams-Russell Cable TV subscribers. If you're a subscriber, you already have everything you need to take advantage of The Cableshop: a phone, a TV set, and your monthly Cable Entertainment Guide.

The Cableshop. A whole new way to go shopping before you go shopping.

Stay Home and Shop Around.

An Issue of Support: Who Pays?

In addition to offering diversity of choice and time and creating an active viewer environment, video of the 1980s differs from television—and, in fact, from all major media—in its source of financial support.

The radio, television, and outdoor communications media all depend solely on advertising revenues for economic survival. Magazines and newspapers require advertising to exist, although a significant contribution to their financial well-being comes from newsstand and subscription revenue.

The new electronic media are quite different. Their economic health is dependent on consumer financial support as reflected in subscription fees and the purchase of a variety of video hardware and software. Advertising has a definite role to play in the economics of the new media, but in many instances it is a supplemental role, adding to their profitability rather than creating it. Individual cable satellite networks and certain interactive services will require advertising support to exist. But cable, video cassette recorders, video disc players and other video media overall will survive and prosper with or without it. If advertisers recognize this, they will perhaps be better able to develop advertising strategy and creative approaches for video of the 1980s.

All of the major broadcast and print media will be impacted by the new electronic media. We will focus on this in the next chapter.

CHAPTER 4

The Impact of the "New" Media on the "Old" Media

Since the word *cable* is generally followed by the word *television*, many advertisers tend to regard it as merely an extension of TV—like network television, spot television, and, therefore, "cable television." In its earliest days, when it served only to improve reception and bring television to areas that otherwise could not receive the signals, cable was in fact merely an extension of television. Today, however, cable has a range of programming and services all its own.

Cable is not simply television with smaller numbers than network and spot TV. It is a new and distinct communications medium that offers high viewer involvement and selectivity and is, in many ways, closer even to many other media than it is to television. Cable is cable—or cable video, if you will. And just as it is important to consider its impact upon television, it is equally important to consider its potential effect on radio, newspapers, magazines, and out-of-home media.

Television

The moment an individual has the capability of receiving cable television, pay television (either cable or over-the-air), or has some form of video recorder or playback device, his conventional television viewing patterns become altered.

Audience Levels

Nielsen studies have found that cable, particularly pay cable, leads to increased television usage, but network viewing levels are cut into among these homes. In addition, among homes with any of the "time-shift" media, viewing patterns are flip-flopped in a variety of ways as consumers buy or rent prerecorded programming or tape broadcast programming to watch at a later hour or date.

The impact of a strong pay movie upon network television audience levels was powerfully demonstrated on November 1, 1983. Home Box Office premiered "An Officer and a Gentleman" to a 39.4 rating and 52 percent share of audience in its estimated universe of 12.5 million subscriber homes. According to the A.C. Nielsen Co., this was the largest HBO movie audience since Nielsen began metering the service. Three network shares of audiences that night fell to 71 percent from 81 percent during earlier weeks of the 1983-84 season.

The new media will affect the advertising and marketing capabilities of television as more and more households become equipped. By the end of this decade:

- Cable television, DBS and other new over-air program transmission services will be in over 60 percent of the nation's homes.
- Video cassettes and discs will be in one-half of the households, letting people watch what and when they want.
- In more than one out of every 10 households, viewers will interact with their sets via teletext and videotex services.
- In over 30 percent of all U.S. homes, the home computer will be a part of everyday life.
- The spread of videocassette recorders and computers will result in an increase in the average number of television sets per household from 2.3 to three.

Since viewers will be doing more and different things with their television sets, the levels of homes using television will increase. In prime-time hours, for example, homes using television will increase from 59 percent in 1981 to 62 percent by 1990. Network shares of audience will, however, decline from 83 percent to 60 percent. Thus, network television will remain the most mass of mass media, but the average program rating will have dropped from a 16 to a 12 by the end of the decade.

Of equal importance is the fact that the declines will not be uniform across all groups, programs, and times of the year. Access to the new media can be expected to climb much more rapidly among the better educated, more affluent, younger and more innovative families. These are the most desirable advertising prospects. Live television—sports, news, and the most popular shows—will remain strong. "Marginal" shows—what Paul L. Klein, a former executive vice president for programs of NBC-TV referred to as the Least Objectionable Programs—will suffer most. They will not be all that a person has available to view. There will be Most Attractive Alternatives. And, similarly, during the increasingly long "rerun season," there will be new electronic video choices and even greater network audience declines.

A serious question arises as to whether—in the face of declining audience levels and rising program costs—three television networks will each be able to afford the same programming services as they do at present.

While we refer to three network "average" ratings, the number is not the same for each network. When the three networks "average" a 16 rating, each network does not have a 16. And when the average becomes a 12, there would similarly be a range in network delivery. (See Table 4-1.)

TABLE 4-1 Network Ratings—Past and Future

	Average Network	X Network	Y Network	Z Network
Past	16	18	16	14
Future	12	13	13	10

The question is: Can the network with the 10 rating in the future afford the same prime-time program lineup it had in the past, since it probably could not generate the needed advertising revenue to match its rising costs?

Or is it not possible that this network might begin to rely more heavily on less expensive news and nonfiction, live talk, and participation shows and on an increasing number of repeats, which would further impact on its audience levels. This scenario is certainly something that must be considered in the face of new media alternatives.

Commercial Flipping and Zapping

While most television research has focused on how cable will affect broadcast audience levels, perhaps a more important area of concern is how it will affect advertising attentiveness. In buying radio to reach teenagers, it has been noted that when the news comes on a comtemporary music station, kids will quickly flip to the nearest alternative where the rock beat is still blaring forth.

With cable, this same phenomenon will occur as the quantity and variety of programming increases *and* as there are more remote control devices in use. Television viewers know that network movie breaks are about two minutes long. They also know that between the end of one show and the beginning of the next lies about five minutes of advertising and other nonprogram material. The result: when a broadcast commercial break occurs, viewers will be able to quickly shift to CNN for an update on the news, MTV for a video clip, to The Weather Channel, or to a home shopping channel—or to any of a score of other channels with ease and with absolutely no compunction about skipping an advertiser's commercials. Figure 4-1 shows how viewing patterns may change.

"Zapping" is not a totally new, new media phenomenon. Before cable, researchers attempted to study behavior during commercial breaks and the incidence of switching to other channels, leaving

FIGURE 4-1 Commercial Flipping and Zapping

	Movie	Commercials	Movie	

↘ MTV ↗
↘ CNN ↗
↘ ESPN ↗
The Weather Channel
Home Shopping
∞

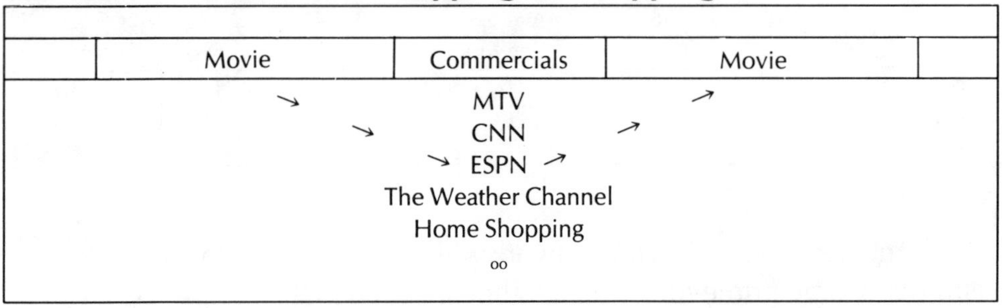

the room or remaining but engaging in activities that might conflict with viewing (like reading) or might not (like eating).

Thus far, however, the research has been far from conclusive. In some cases, it has relied upon what people reported doing. In other cases, it has not studied the various patterns for different lengths of commercial breaks, and, in still others, it has not focused on the full impact of remote devices with new, multiple channel cable systems.

What is known is that the impact of zapping depends on having something to "zap" with and someplace to "zap" to. Today's new remote control devices and cable channels, which cater to short-span viewing, provide both of these. (See Figure 4-2.)

FIGURE 4-2 A MODEL FOR ZAPPING

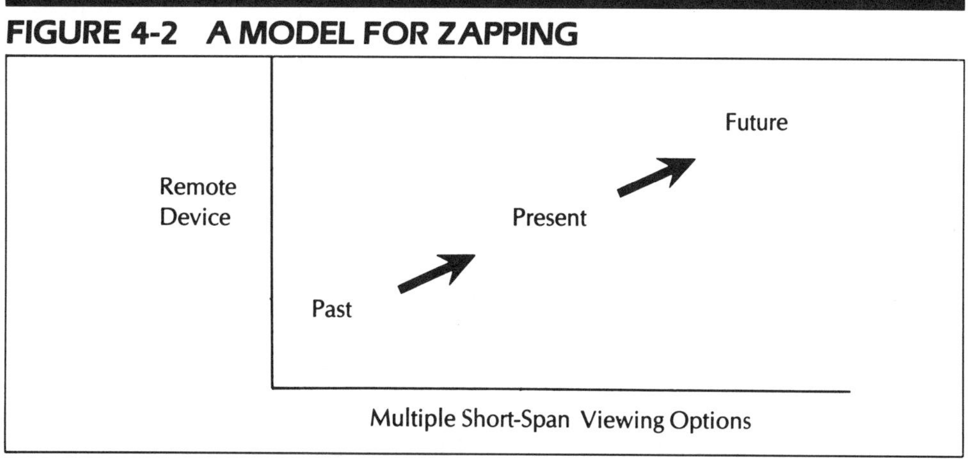

In homes with video cassette recorders, flipping and zapping take on an added twist. For example, viewers can tape the early evening or late news, and then play it back and fast-forward past stories—or commercials—they have little interest in. The viewer chooses what he or she wants to see just as a magazine reader flips past articles and advertising he or she chooses to ignore.

As video cassette recorder ownership climbs, so does the advertiser's concern that it can provide an easy means of editing out his commercials. In its American Consensus Report, *The New Technologies: The View from the Viewer-II*, Benton & Bowles found that 61 percent of VCR owners "always" or "usually" skip commercials when viewing a program they had recorded at an earlier time (May 1983). And in its tracking studies, The Roper Organization reported that 40 percent of the people who recorded a program for later viewing said they edited out all of its commercials and station breaks (December 1983).

If marketers fail to consider these ways in which the new electronic media affect advertising attentiveness, they naively assume that as cable and time shifting increase in impact, television's commercial impact will remain unchanged. (See Figure 4-3.) In fact, the pattern looks more like Figure 4-4. Or, perhaps, if advertisers and broadcasters respond to flipping and zapping by making changes in commercial formats and by making the commercials themselves more interesting to hold viewer tuning and attention, the pattern may even reverse itself and look like Figure 4-5.

FIGURE 4-3 Unchanged Viewing Patterns

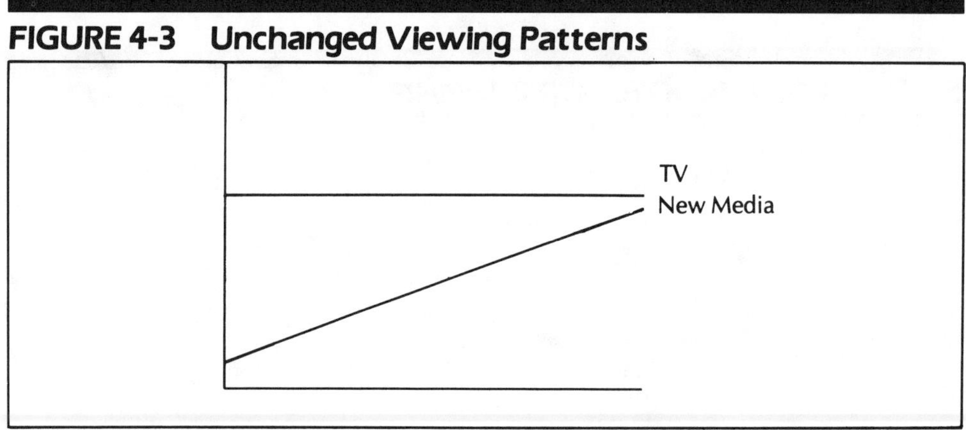

Many videotex marketers have considered how their product might be tied into broadcast television. Viewers would be encouraged to turn from a broadcast commercial to videotex in order to get more information on a product or service. For example, a food manufacturer might offer detailed food preparation suggestions. An insurance company might suggest that viewers take a home safety quiz on their videotex channels.

While this "cross-media marketing" obviously would benefit the advertiser engaging in it, the reaction of other advertisers to the suggestion that viewers turn from the channel they are watching would be far less than positive. It is questionable whether or not the broadcast networks and stations would ever clear commercials that suggested zapping an ongoing program.

FIGURE 4-4 New Viewing Patterns I

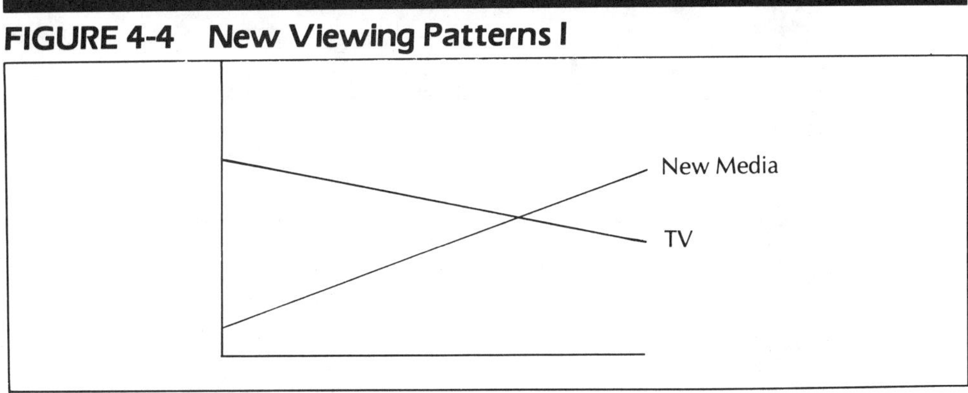

Zap Proofing the Advertising

Despite the problems associated with zapping, advertisers are beginning to formulate ways to minimize its effects—and, like all advertising, they involve better creative approaches and a little imagination. In late 1984 General Foods began a unique cable strategy that involved a series of two-minute entertainment-oriented spots on several cable networks. The commercials combined 90 seconds of helpful kitchen tips with a product message for one of GF's many food products. This blending of entertainment and advertising may be one way to cut down on commercial zapping.

Another effective "zap-proof" creative technique is the "mini quiz," such as J. Walter Thompson USA developed for Dunlop. The first part of the message showed a sports event and asked a question. A Dunlop golf or tennis ball commercial followed, and the spot concluded with the answer to the question. This creative approach re-

FIGURE 4-5 New Viewing Patterns II

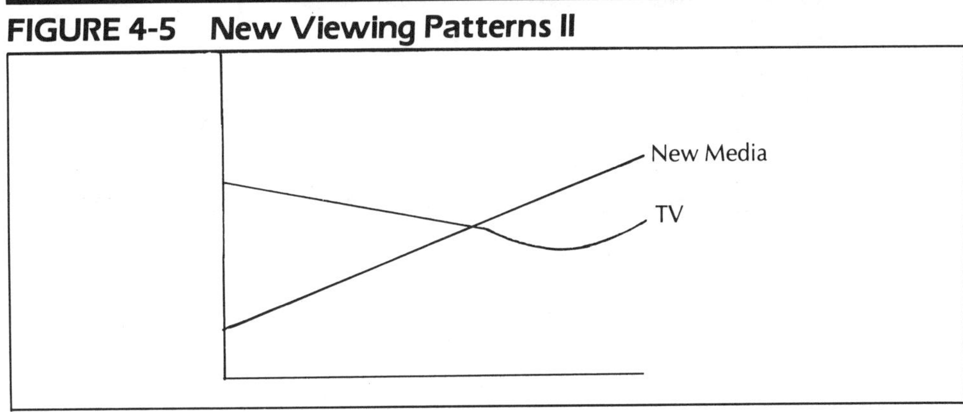

Using print to support a cable effort, General Foods promotes "Shortcuts" with this ad in *The Cable Guide*.

THURSDAY, OCTOBER 24

1:30 AM TO 7:00 PM

Time	Ch.	Network	Program
1:30	24	USA	*Japan Today*
2AM	31	ESPN	*Inside The PGA Tour*
	24	USA	*College Basketball: Syracuse vs. St. John's*
2:05	18	SHO	***PRETTY MAIDS ALL IN A ROW*** Nudity, profanity (R-1:31) p.46
2:15	29	A&E	*To Be Announced*
2:25	14	HBO	***HISTORY OF PRO FOOTBALL*** (1:29) p.45
2:30	29	A&E	*At The Met: Voyages—The Journey Of The Magi*
	31	ESPN	*SportsCenter*
3AM	29	A&E	*The Romantic Spirit*
	31	ESPN	*SportsLook*
3:20	8	WTBS	*Movie: Purple Taxi* (1977) Drama
3:30	31	ESPN	*Top Rank Boxing*
3:40	18	SHO	***10 TO MIDNIGHT*** Violence, profanity (R-1:41) p.47
4AM	14	HBO	***DISPOSABLE HEROES: AMERICA UNDERCOVER*** (1:00) p.30
	24	USA	*College Basketball: Memphis St. vs. Florida St.*
5AM	14	HBO	***THE LONELY LADY*** Nudity, violence (R-1:31) p.46
5:30	18	SHO	***FAERIE TALE THEATRE: JACK AND THE BEANSTALK*** (NR-:52) p.31**

24 THURSDAY OCTOBER

Time	Ch.	Network	Program
6AM	31	ESPN	*Business Times*
	24	USA	*BizNet News Today*
6:30	14	HBO	***THE GREAT WHALES*** (:59) p.33
	18	SHO	***FAERIE TALE THEATRE: SNOW WHITE & SEVEN DWARFS*** (NR-:53) p.31**
7AM	30	NICK	*The Adventures Of Black Beauty*
	24	USA	*USA Cartoon Express*
7:30	14	HBO	***EMMA AND GRANDPA: WINTER*** (NR-:30) p.33
	18	SHO	***KAVIK THE WOLF DOG*** (NR-1:39) p.32
	30	NICK	*Lassie*
8AM	14	HBO	***THE YEAR OF LIVING DANGEROUSLY*** Violence (PG-1:54) p.47
	29	A&E	*Rising Damp*
	30	NICK	*Belle And Sebastian*
8:30	29	A&E	*The Fainthearted Feminist*
	30	NICK	*Today's Special*
9AM	29	A&E	*Women In Jazz*
	31	ESPN	*SportsCenter*
	30	NICK	*Pinwheel*
	24	USA	*Calliope*
9:05	8	WTBS	*Movie: Angel Baby* (1961) Drama
9:30	29	A&E	*Harvest Jazz*
	31	ESPN	*SportsLook*
10AM	14	HBO	***THE LUCKY STAR*** Profanity (PG-1:45) p.46
	18	SHO	***THE GREAT SINNER*** (NR-1:50) p.45
	31	ESPN	*College Basketball: Ohio State vs. Illinois*
	24	USA	*Sonya*
10:30	29	A&E	*The Hot Shoe Show*
11AM	29	A&E	*Yes, Minister*
	24	USA	*Great American Homemaker*
11:30	29	A&E	*Last Of The Summer Wine*
	24	USA	*Peyton Place*

CABLE'S TASTIEST TWO MINUTES.

"SHORTCUTS"™

CREATIVE FOOD IDEAS FOR BUSY PEOPLE:

- SHOPPING TIPS
- NUTRITION
- ENTERTAINING
- QUICK & SIMPLE RECIPES

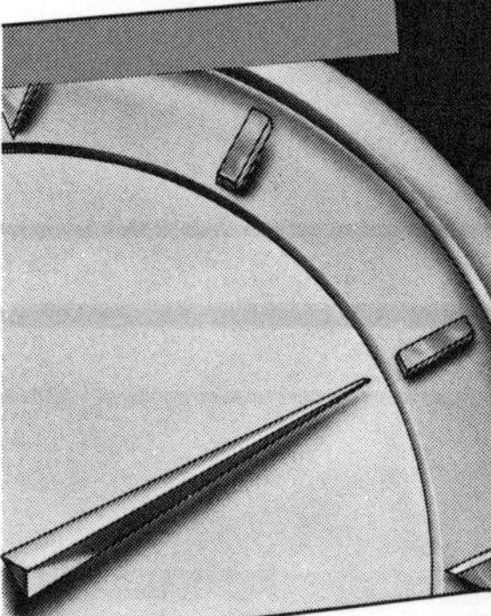

SPONSORED BY GENERAL FOODS

EVENINGS ON:

CBN, USA, NASHVILLE, LIFETIME & CNN

The Cable Guide

October 1984

lated the product to the program and also was able to hold the viewer's attention through the advertising message.

Radio

Like radio, cable is a highly localized medium. In fact, it is even more sharply localized. A cable system's boundaries can be precisely defined, and, unlike broadcasting, they are usually smaller and more manageable in size. Because of this, cable offers advertisers a new electronic media form in communities where they have relied mainly on radio because television was either unavailable or unaffordable. Cablecasters in some markets are even considering joining forces with local radio stations in videocasting popular talk radio shows or disc jockeys. Sight would be added to sound in a "reverse simulcast."

Cable can provide many advertisers with the same low-cost, high-frequency, target-audience delivery they may now find in radio but not in television. For example, MTV: Music Television offers 24 hours a day of *video* music in *stereo*. It's a blend of TV and FM stereo radio that may feature artists singing their songs—or acting them out. Sometimes they are animated sometimes they use experimental video art.

Advertisers who now use contemporary radio to communicate with youth or young adult target markets can deliver the same audience via MTV on both a national network and (through local availabilities) on a local community basis. Chevrolet, for example, ran the first long-form, 90-second stereo commercial on MTV for Camaro. It was an impressionistic montage set to a soft rock tune. A series of fast clips included body builders, a street sign reading Z28 (the car's model number), and a shot of the Camaro cruising past a desert sunset.

Since its birth on August 1, 1981, MTV has exhibited enormous growth in both subscriptions and viewership. Other basic and pay cable networks and broadcast networks and stations recognized its uniqueness and its ability to communicate with the very hard-to-reach 12-34 year old audience. The result has been the creation of over a dozen video music services providing advertisers with many opportunities to reach what formerly might have been an exclusive rock radio audience. These include:

- MTV: Music Television
- Night Tracks (WTBS)
- Radio 1990 (USA)
- Night Flight (USA)
- Friday Night Videos (NBC)
- MusicChannel (SPN)
- Video Sound (BET)
- Video Music Countdown (Washington)
- Video Rock (Philadelphia and Erie)
- Rock Street (Oklahoma City)
- Miscellania (Nashville)
- 36 Juke Box Video (Atlanta)
- FMVision (Topeka)

In January 1985, MTV started a second 24-hour music service—VH-1—aimed at 25 to 49 year olds. And The Disney Channel now features music videos built around their animated cartoon characters.

Cable has taken a leaf from the talk radio programmer's handbook. Half a dozen cable networks are all featuring live call-in shows where viewers can voice their opinions on politics, religion, music, money and sex. What we have is a new program category—"Talk Television."

In still another area, cable satellite transmission is bringing new radio stations to markets that formerly did not receive them. Chicago's classical FM station—WFMT—is transmitted across the country as a "Super (radio) station," the radio equivalent of Ted Turner's Atlanta Superstation (WTBS), WGN (Chicago), and WOR (New York).

Cable is attracting a growing number of 24-hour, satellite audio channels. Offering commercial-free jazz, country, big band/nostalgia, classical, easy listening, mellow rock oldies, and top 40 music, the channels transmit to subscribers who have home FM receivers with converters. As cable penetration and the growth of these audio services increases, the audience levels of commercial radio stations may suffer.

Newspapers

In many small communities and suburban areas, the only local medium is the suburban or shopping newspaper. Cable video will change this, providing a new selling medium for retailers, local dealers, and franchises. For the people in the community who have depended exclusively on their local paper, cable will deliver a wide array of community news:

- programs for and about local schools and senior citizens.
- programming from local business and civic organizations.
- coverage of village board meetings.
- programming from the community library and local colleges.
- coverage of local athletic events, celebrations, parades, and festivals.

All of these offer advertising potential, and recognizing that cable could cut into their local revenues, astute newspapers aren't playing the "ostrich with its head in the sand" but are quickly recognizing the value of the old adage. "If you can't lick 'em, join 'em!" One company is Leader Tele-Cable, which publishes the *Eau Claire (Wisconsin) Leader Telegram.* Not only does the company publish an electronic version of its daily, but it also has developed a classified ad channel that offers a full-color picture of an item for sale along with the price and seller's phone number. It complements the newspaper, but in an entirely different advertising form. We can also expect to see newspapers creating and cablecasting the news directly from their editorial offices, perhaps even joining together to provide a national satellite service.

One of the more imaginative alternatives to traditional newspaper advertising was developed by Televised Real Estate, Inc. Since 1979, it has occupied one channel on a local cable system in Spokane, Washington, and in September 1981 it began cablecasting on two systems in southern Orange County, California. For 17 hours a day, this real estate channel focuses on a variety of industry-related topics, but its main thrust is 60- and 90-second "commercials" for homes, clustered by communities. In each ad, the property is quickly toured and described, with special features highlighted. The end of each ad includes price and financing. Commercials for new home developments run to "infomercial" length of three to eight minutes. The real estate channel competes with the traditional newspaper real estate listings and can provide the incentive for a follow-up visit to the home.

Among advertising's dreariest categories is the $75 million plus that is spent annually for "Tombstone Ads." Placed by investment banking syndicates to announce new issues of stocks and bonds, there is no sales pitch, simply a black-and-white listing of the issuer, number of shares, price per share, and the underwriters. Traditionally confined to placements in the financial press and major city newspapers, *Business Times*—the morning business news on ESPN—has actively solicited tombstone ads. With the creative use of a character generator, the layouts can easily and inexpensively be translated from printed page to video screen.

Magazines

Magazines are embracing the new media avidly. Home Box Office has carried *Consumer Reports, Ms., Money,* and *Sports Illustrated* programming. *Better Homes and Gardens, Family Circle,* and *Scholastic* have developed cable shows, and *Playboy* programs an entire adult channel.

As the number of cable channels expands in the years ahead, viewers will find themselves browsing through video selections just as they now browse through the almost infinite variety of magazines at the newsstand. Already today, cable services are devoted to:

- Arts and Culture
- Blacks
- Business and Finance
- Children
- Music
- News
- Religion
- Spanish

- Consumers and Shopping
- Country
- Health
- Sports
- Weather
- Women

Ahead may be advertiser-supported services focusing on magazine categories dealing with:

- Automobiles
- Beauty and Fashion
- Cooking and Foods
- Farming
- Gossip
- House and Home
- Literature
- Nature
- Senior Citizens
- Travel

One new media creation of 1984 was the "Fashion Video" which weaves together beautiful people, high fashion and the music of today. It is even possible to imagine fashion videos which promote consumer products (Ivory soap), or are used between or during intermissions on HBO or Showtime. There are opportunities here not only for advertisers but for the women's fashion magazines to develop fashion videos.

Many publications are working to develop videopublishing—programming for cassette and disc. These programs would encompass not only the consumer field but also could be developed to cover many business, industrial, and trade publications. Financial support would come from both subscriber and advertiser revenue.

A good example of how videopublishing can reach a small, highly specialized audience is the *"Advideo Journal,"* published by Charles Mandel, former publisher of *Science Digest. "Advideo Journal"* uses documentary-style videotapes to report on significant stories of interest to advertisers and advertising agencies. Revenue comes both from subscribers and from advertisers such as *Atlantic, Business Week, The New Yorker, TV Guide,* and W.R. Grace & Co. who run commercials in the program.

Used for news, reference, seminars and training, each issue (tape) shows and tells about a variety of major advertising topics. Examples in the first three issues were:

- Developing the Diet Coke Campaign

- The State-of-the-Art in TV Animation
- The Entry of Kronenbourg Beer in the U.S.
- How Ogilvy & Mather Landed the $40 Million AT&T Account
- The Growth of Special Events Sponsorship
- The Role of Advertising in Chrysler's Turnaround
- A Selection of 1983 International TV Clio Award Winners.

In Los Angeles, an advertiser-supported "video magazine" aimed at travelers is beamed into hotel rooms and also picked up by several cable systems that transmit it into subscriber homes. The 30-minute program, "Keys to the City," is played around the clock and is available in English, Spanish, German, French, Japanese, and in sign language. A typical TV magazine format with a host and hostess, the show features public service segments, features on local culture and samplings of entertainment attractions and goods and services.

An effective twist on videopublishing was the "Reader's Digest Do-It-Yourself Show" on the USA Cable Network. It was a 13-week series of entertaining and informative visual companion pieces to the "Reader's Digest Complete Do-It-Yourself Manual" and "Fix-It-Yourself Manual." A husband-and-wife team demonstrated how to build and how to fix things. The entire sequence of steps in each project was illustrated in color with computer-animated graphics.

It is not unusual to anticipate that the new electronic media will impact upon the magazine industry and that print will seek ways to interact with the new video forms. After all, magazines, cable, VCRs, and video discs have in common the attraction of an affluent and educated audience. The implication of this for advertisers is significant. Many marketers with highly upscale products have found that the audience appeal of broadcast television is too broad and have confined their advertising efforts exclusively to upscale magazines. Special-interest cable programs can extend these marketers' media options considerably and allow them to reach upscale audiences with video just as they do now with special-interest magazines.

One example is the increasing interest the public has shown in business and financial news, which has resulted in an expansion of business coverage in newspapers as well as the growth of national, local, and regional business publications. These, in turn, have been joined by a growing number of television and cable programs aimed at the same target:

- BizNet News, Early Edition (USA Cable Network)
- BizNet News Today (Modern Satellite Network)
- Business Times (ESPN)
- Business View (Modern Satellite Network)
- Dow Jones Cable News (Cable)
- Financial News Network (Cable and Syndication)
- Inside Business (Cable News Network)
- Insider Business Today (PBS)
- It's Your Business (Syndication)
- Moneyline (Cable News Network)
- Moneyweek (Cable News Network)
- Nightly Business Report (PBS)
- Pinnacle (Cable News Network)
- Taking Advantage (Syndication)
- Wall Street Journal Report (Syndication)
- Wall Street Week (PBS)
- Weekend Business Report (The Learning Channel)
- Your Money (Cable News Network)

Advertisers seeking to reach a better-educated, upper-income audience through financial publications can add the video versions of these publications as a very logical market expansion opportunity.

"Business Times," cable's weekday morning business and financial news show on ESPN, makes a direct comparison of its delivery with that of *Business Week* magazine.

- For $34,000, an advertiser in early 1984 could buy one four-color page in *Business Week* with a total U.S. circulation of 800,000

- For less than $30,000, the same advertiser could purchase one 30-second commercial a day—five days a week—for four weeks on "Business Times." These 20 messages would reach over 900,000 different homes an average of 1.7 times each.

While a comparison such as this does mix apples and oranges, it makes the point of cable's ability to provide exposure of a business message in a business environment with multiple exposure at costs competitive to business magazines. It is a strong selling tool.

Advertisers can also use cable to deliver a "unique editorial ruboff," just as they often do today in special-interest magazines. It is quite common for an advertiser to seek publications that have editorial subject matter that can reinforce its product message. Thus, a furniture manufacturer looks for magazines with home decorating and remodeling features, while a clothing manufacturer prefers to locate ads near fashion editorials. Such opportunities are available in cable, often in some very unusual ways.

A weekly series on the CBN Satellite Network is "Fresh Ideas with Claire Thornton." Each week's show includes a report on what produce is in season and is plentiful and a feature focusing on a particular aspect of the produce industry, such as a grower, a commission, or an association. Finally, Claire provides tips on how to select, store, and prepare fresh produce.

Several years ago, a leading salad dressing manufacturer examined magazines not only on the basis of the amount of food editorial they carried but also on the basis of the number of editorial items dealing with fresh vegetables and salads. "Fresh Ideas with Claire Thornton" provides the same kind of editorial environment normally confined to print. It is not unusual to find the show's advertisers including companies such as Sunkist, Dole Pineapples & Bananas, Washington State Apples, Kraft, and the Imported Winter Grape Association.

A number of programs have been developed for Lifetime, CBN, SPN, and other cable networks to focus on baby and child care, just as special interest magazines do. "The American Baby Cable Show," "Mother's Day," "Infants and Toddlers," "Pediatrics and Parenting," and "Bringing Up Kids" all offer advertisers interested in the infant market the targeted appeals of such magazines coupled with the video impact of cable. While audiences are small, they contain a much higher concentration of mothers and mothers-to-be than would be found in general daytime shows in broadcast television. And because the viewer actively seeks out such programs to watch, it would be logical to assume that she would be highly responsive to its advertisers.

One measure of the audience attraction for these shows involved the offer of a *free* one-hour videotape cassette, "The Best of American Baby TV Show," to childbirth educators, who averaged nine classes a year. Many taught in hospitals and institutions where other instructors also used the cassette in their classrooms. Over 1,300 requests were received for the free cassette, which included the advertising spots of the show's sponsors for added exposure.

The SI/ESPN Package was created in mid-1981 as the industry's first magazine/cable combination buy. Since *Sports Illustrated* and ESPN delivered similar subject matter to a similar upscale, educated male audience, the package offered advertisers an innovative way to use the two media in combination.

Robert L. Durkee, Director of New Business at ESPN, reported that 36 out of 130 sales proposals made during the first two and one-half years of the plan resulted in sales. Since the program doesn't offer any rate discounts and negotiations are conducted separately with SI and ESPN, part of the plan's appeal is that advertisers earn a 3 percent merchandising credit. It might be used to have a professional golfer play with a company's top customers, bring a sports star to a sales meeting, or for any other sports marketing effort.

A similar cable-magazine partnership was established in 1984 between Financial News Network and Financial World Magazine. The aim is to target the upscale audience that both have in common—corporate image advertisers, financial service companies, and high end consumer goods manufacturers.

Out-of-Home

No two media are as similar in geographic concentration of audience as are out-of-home and the new video. Out-of-home can profit from this. Outdoor locations within the community can specifically highlight the value of signing up for cable. By picking sites near theaters, pay services, and home video manufacturers can reach people on their way to the movies and highlight the economy, convenience, and comfort of watching films in one's own home. Transit cards can include tear-off schedules of pay service offerings which include mail-in cards for further information.

The impact of the array of new media devices can actually create important new areas of revenue growth for out-of-home.

New Media to Reach the New Media Industry

The focus of most new electronic media has been to deliver entertainment and information to the general consumer. Instead of reaching out to the general public, however, several services are designed to communicate within the $100 billion electronic media industry itself.

One of these, Videotex Information Corp., developed "VideoLog" to provide updated information to computer literate people in the industry who are looking for data that can be rapidly updated. Advertisers used to buying pages in electronics trade magazines buy "electronic pages" in "VideoLog" that run for a year. The "readers" or "subscribers" pay a small data-transmission fee and receive the information through an add-on delivery on their personal computers.

Also aiming at a similar target industry market is "Cable Convention News." Many trade publications provide special show "extras," which are dropped under the hotel room doors of attendees in the middle of the night. "Cable Convention News" uses videotape and closed circuit television to transmit convention activities and news directly into hotel rooms. Updated daily, it features panel session reports, interviews with convention attendees and exhibitors, new product demonstrations, and comments from the industry's key decision makers. Advertisers support the service using either existing commercials or messages specifically prepared for the program.

Planning Ahead

What all of this means, in brief, is that the world of media will never be quite the same. It will provide new opportunities, new challenges, and new problems. Planning, in particular, will be radically different from planning in the past. That is the subject we will now turn to.

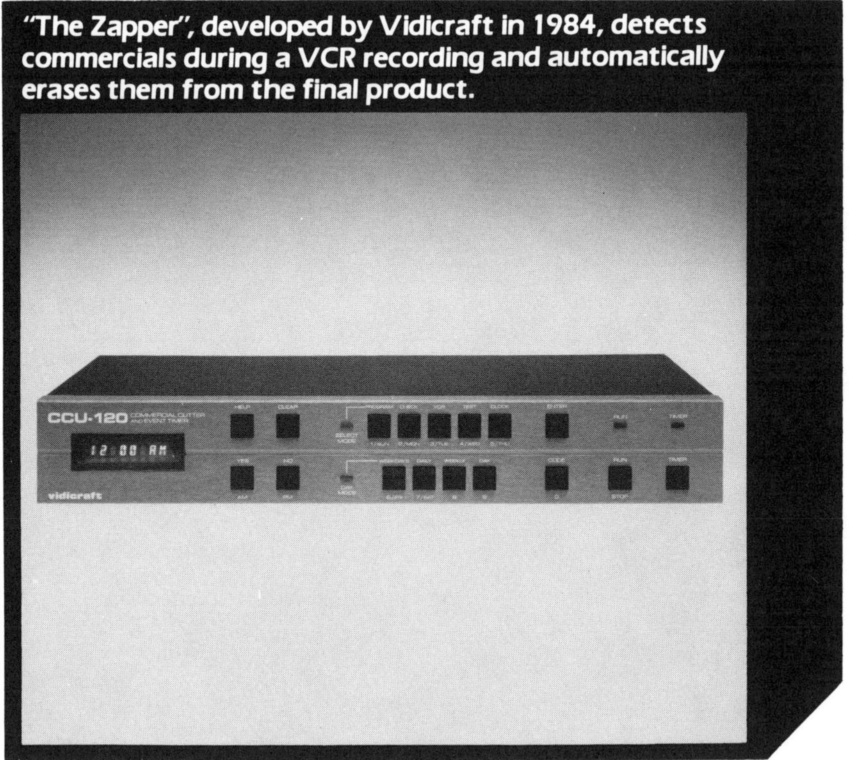

"The Zapper", developed by Vidicraft in 1984, detects commercials during a VCR recording and automatically erases them from the final product.

CHAPTER 5

The New Media Planning Process

The planning and execution of cable advertising strategy can be described as a classic case of trying to put the right round pegs in the right round holes in an environment where there is an equal number of square holes confronting the advertiser. The new era of video diversity presents as many problems as it offers opportunities:

- How does the advertiser know *where* to reach the best prospects?
- How does the advertiser know *how many* have been reached?
- How does this advertiser know what is the *cost* of reaching prospects, if indeed they can be reached?
- How does an advertiser know *what kind* of advertising can communicate most *effectively* with them?

In short, how does an advertiser piece together a cable plan when everyone is doing his or her own thing with his or her own television screen?

Having learned everything he can about all of the new cable video opportunities that are opening up, what does the advertiser do *now?*

The secret to success is: *Never evaluate a cable opportunity in isolation!*

Cable video is a medium that should be evaluated in the context of *all* of the media options available to an advertiser. (See the configuration in Figure 5-1.) The decision to use cable must be based on the fact that it offers a new or better way to accomplish an advertiser's marketing and/or communications goals.

FIGURE 5-1 Cable Video—One of Many Advertising Options

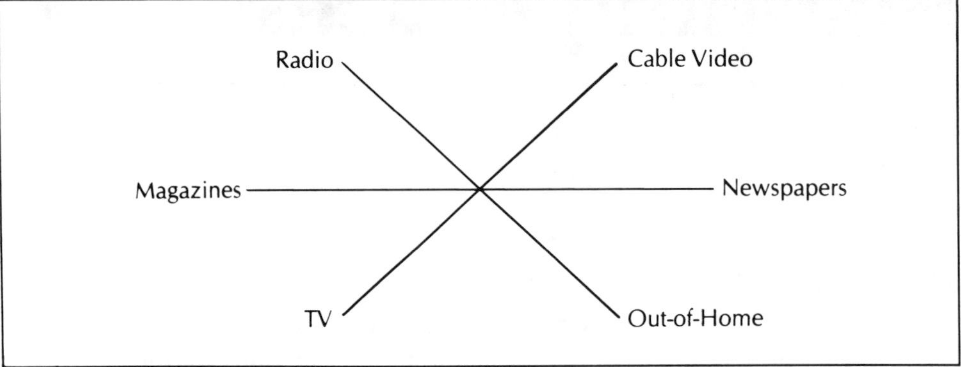

What Can Cable Do for You?

Before deciding to use cable, an advertiser must carefully evaluate his existing media plans to determine how well they are accomplishing everything he would like them to do. An effective way to do this is to take the Advertising Effectiveness Quiz:

- I am basically satisfied with my advertising and with my media plan.
- It appears to showcase my product (or service) well.
- Research shows that it is conveying to my customers pretty much what I want to tell them.
- It seems to be efficient.
- I believe that it is doing an effective job.

I only wish that (fill in the blanks)

____

____

____

____

____

____

The answers will take a variety of forms, including:

- I only wish the media conveyed my messages in a more compatible environment and acted as more than just a commercial "carrier."
- I wish I could afford to sponsor something where I could get some real identity.
- I wish I had more time to tell my story. Thirty seconds is not enough, but that's all my television budget can afford.
- I wish I could have the advantages of television, but with the ability to say everything I now say in my print ads.
- I wish I could advertise somewhere that appealed more to the very special people that buy my product.
- I wish I could afford the high frequency of exposure I need but can now get only in radio.
- I wish I could reach those upscale cable homes more frequently.
- I wish I could zero-in more closely on some of the small towns where my dealers are located.
- I wish I could run a dozen or more different commercials, but I can't afford to spend all the money it takes to produce them for broadcast television.
- I wish I could experiment a bit and explore some new commercial ideas—maybe even try some direct response options.

Examining these responses can help an advertiser decide whether or not cable can provide a positive addition to the media mix.

Looked at in the context of all media, cable offers a variety of highly *specialized programming* that allows advertisers to zero-in on highly *targeted audiences* they might otherwise find hard to reach. Cable advertising is available at relatively *low costs per advertising unit.* This allows marketers who cannot afford broadcast television to take advantage of its visual and audio communications potential. Cable has the ability to provide *high frequency of exposure* to an advertiser's prospects throughout the day and offers *flexible message lengths* that meet advertiser communications needs rather than broadcast time limitations. Cable offers *program sponsorship opportunities* with advertiser identity. The program becomes more than just a commercial carrier. Cable provides *product or service exclusivity* in programs at affordable out-of-pocket expenditures. Cable audiences, in general, are *better educated, above average in income, employed in higher-level jobs, and younger.* Cable offers *opportunities to test creative ideas* at very low media costs. Cable provides many opportunities to tie-in advertising messages with *direct response offers.* Cable is *highly localized* and can provide advertising support for franchisees, dealer organizations, and wholesale and retail sales forces.

Looked at another way, cable offers: the visualization of television, the targeted formats and specialization of magazines, the low unit costs and frequency of radio, and the local interest and appeals of newspapers. (See Figure 5-2.)

FIGURE 5-2 Opportunities Offered by Cable

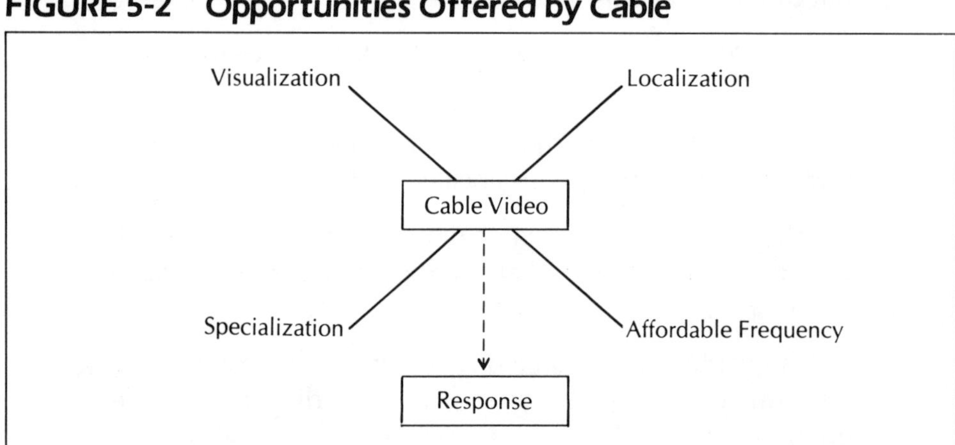

Understanding Cable's Deficiencies

Cable is such a new medium that considerable judgment is required in evaluating how to use it most effectively. While cable has many assets, many of these are, at present, relatively unproven. Thus, in the mid-1980s, cable as an advertising medium is perhaps at the same stage as broadcast television was in the mid-1950s.

Lack of Audience Information

A giant missing link in the cable advertising evaluation equation is a substantial body of audience research. Most people recognize that cable audiences are substantially smaller than broadcast audiences. However, the information that is available today comes in a variety of forms and a number of different bits and pieces. Furthermore, there is not full agreement within the industry as to just what to measure and how to measure it (more on that in Chapter 8).

Even with the lack of data, however, the cable industry has recognized the need to fill the void. When the first edition of this book was published in 1982, only HBO, WTBS, and the Cable News Network had continuous and consistent metered audience data available from the A.C. Nielsen Company. By early 1985, 11 services had joined the group. For other cable networks and most local systems, such information is available as occasionally ordered special analyses.

Spotty Coverage

In some major markets, cable's coverage is still very spotty. For example, in November 1984, 44 percent of the homes in the country were hooked up to cable. However, in Chicago, Washington, D.C., and Minneapolis-St. Paul, less than 25 percent of the homes had cable. (See Table 5-1.)

For some markets in the process of "cabling up," the spotty coverage may reflect a low level of central city penetration (or none at all) but high (or growing) cable coverage in surrounding suburbs. Advertisers should examine these patterns closely since for some products (snow blowers, lawn care equipment, home maintenance items, etc.), the coverage he gets may be extremely desirable.

Frustrations in Evaluating, Buying, and Creating Advertising

For many advertisers and advertising agencies, the multitude of cable offerings, uncertainties as to how to evaluate them, and frustration as to how to prepare special advertising for them have resulted in considerable confusion and anxiety. Agencies on a traditional 15 percent commission system are also finding themselves in a severe cost squeeze.

TABLE 5-1 Cable Penetration in the Top 20 U.S. TV Markets (Percent)

Market	Percent
New York	34.0
Los Angeles, Palm Springs	34.1
Chicago	23.0
Philadelphia	44.1
San Francisco-Oakland	47.1
Boston, Manchester, Worcester	40.2
Detroit	31.3
Washington, D.C.	22.3
Dallas-Ft. Worth	36.7
Cleveland, Akron	43.4
Houston	42.1
Pittsburgh	58.9
Miami-Ft. Lauderdale	41.7
Minneapolis-St. Paul	19.5
Atlanta	40.0
Seattle-Tacoma	50.6
Tampa-St. Petersburg, Sarasota	42.6
St. Louis	31.6
Denver	37.7
Sacramento-Stockton	31.9

Source: A. C. Nielsen, November 1984

The small dollar amounts involved in most of today's cable buys make it impossible for agencies to profit from them. For example, one network commercial on the 1985 Super Bowl cost $525,000. For this much money, a *very large* cable network *schedule* spanning many months could have been purchased. In fact, during 1984 one advertiser purchased over 500 commercials on *six* satellite networks for $500,000. The cable buys were obviously several times more complex, costly, and time consuming to administer than the Superbowl buy. It is no wonder that agencies may find it necessary to ask advertisers for fees so they can afford to spend the necessary time to negotiate cable buys and produce special commercials for them.

Why Get Involved Now?

With cable's currently small potential audience size and the problems involved in evaluating, buying, and creating advertising for it, many advertisers ask, "Why not wait?"

The answer involves the "risk/reward relationship." Every informed source indicates that by the end of this decade, traditional television will have been transformed into a multitude of video offerings available to a majority of the population. The investment of time and money to explore, experience, and experiment today will better prepare the advertiser to effectively use these new media offerings tomorrow.

Those early pioneers who were attracted to broadcast television during the late 1940s learned what worked and what didn't work at a time when making a mistake carried a very small price tag. Thus, Kraft was the first advertiser to sponsor a weekly television drama, starting on May 7, 1947, in New York when there were only 32,000 TV homes in the entire city. Of course, what was learned then about producing a weekly program far outweighed the costs. Later, in 1956, they began producing all of their commercials in color, even though there were fewer than 200,000 color sets in the country. Again, Kraft knew that food should be shown in color and that color would eventually reach the majority of homes nationwide. The cost of learning how to use color properly in 1956 was far smaller than it would have been if they had waited until years later, when the networks all began to telecast in color. The same holds true with cable. The rewards of learning how to best use cable today far outweigh the risks of procrastinating.

But the potential of cable is not just in the future. Many opportunities exist to use cable efficiently and effectively today. The key is to select alternatives and then implement those that will best achieve a set of advertising objectives, whether they be for the Ford Motor Company or for a Ford dealer in Little Rock, Arkansas, for American Express or for a small bank in the suburbs of St. Louis.

Taking Advantage of Publicity and Promotion

Through the creative use of well-planned and well-timed publicity efforts, the impact of today's buys can be increased and considerably extended. Publicity can promote the cable effort among cable subscribers to increase viewership and announce the advertiser's use of cable to the trade, business leaders, government and community leaders, educators, the press, and other important influences affecting the advertiser's business.

Since many people will not have access to cable, the judicious use of print advertising can announce a cable campaign to many who would otherwise be unaware of it. For example, an ad in the *Wall Street Journal* announced that Kraft had signed up as the first advertiser on CBS Cable. It reached an important group of thought-leaders who would not otherwise have been aware that Kraft Music Hall was returning—this time to cable.

An extremely creative use of cable to cross-promote an advertiser's "noncable" media efforts was developed by The Financial News Network. During an advertiser's flight on FNN, the network will tag commercials with a mention of his other media activities, including network radio and television, magazines, print supplements, catalogs, and national mailings. For example:

- "Watch the Summer Olympics on ABC, sponsored in part by (advertiser), the official (product) of the Summer Olympics."
- "Look for (advertiser) 18-page supplement on how to buy a computer in this Sunday's newspapers."
- "Read all about (advertiser) new mobile telephone system in Friday's Wall Street Journal."

Financial News Network produces the commercial tag using the advertiser's logo slide, the event's logo slide and a voice-over.

At a retail level, MTV has greatly helped the record industry as a promotional as well as a media tool. Retailers can buy generic posters, MTV buttons, bumper stickers, and other materials. A national information phone line also gives retailers the most up-to-date information on the newest video and upcoming MTV concerts so retailers can know which records to stock and promote. Some record retailers are even bringing MTV right into their stores (either live or on tape) where it can be seen on monitors at the point of sale. This is

This ad announced the return of Kraft Music Hall, this time on CBS Cable.

perhaps the first use of a video medium as an "on-air" point of purchase display.

If this sounds avante garde or especially innovative, it isn't—at least not insofar as the shape of the future is concerned. Richard Hagle, marketing director of the *Sales Promotion Monitor*, contends that the entire range of sales promotion devices eventually will be delivered electronically. By mid-1984, three companies—the Electronic Advertising Network, Nuvatec, and Compucard—already had various devices in test market. The EAN system is one example. In-store machines show consumers a variety of promotions at the front of the store. The consumer activates the machine with a credit card and, via a touch-sensitive screen, selects the most attractive offer and then purchases it in the store. The machines were expected to go national in late 1984. Similarly with Compucard: consumers are shown information on products and then can buy them in the store or by using their credit cards. It is only a short step, Hagle points out, to being able to deliver such offers via cable. In fact, Compucard already made some of its offerings via cable in the fall of 1984. Company representatives anticipated being able to make such offers available on a large scale by 1990.

In early 1984, the Silent Network—delivering programming for the deaf and hearing-impaired—was launched to over five million cable subscribers. Advertiser-supported, programs are produced in sign language, voice and open captions. There are obviously strong goodwill, public relations benefits from an advertiser's support of this network.

It is impossible to spell out in detail all of the guidelines for effective publicity; all of the traditional approaches certainly apply. Not only can a cable program be promoted by an effective publicity effort, but publicity efforts themselves can often be turned into well-executed cable advertising programs. For example, many high schools across the country offer evening how-to-fix-it courses in conjunction with a local hardware store or lumber dealer. A natural extension of this would be the production of a series of how-to programs by the community's local cable system. These would be funded by the participting hardware store or lumber dealer and would feature their products in use. Thus, what was simply a public relations effort could become an effective cable advertising program.

A very successful prototype for this kind of program is already in existence. The Public Broadcasting System's "This Old House," which originates from WGBH in Boston, is devoted to the how-to approach. Specifically, the program shows how to handle nearly every conceivable remodeling problem so that an old house can be completely refurbished and made ready for sale or habitation. A series of books based on the program also has been produced.

Establishing Cable Advertising Objectives

Effective cable advertising executions begin first by answering the same series of questions you answer in evaluating magazines, newspapers, out-of-home, radio, and television:

- *Who* are the people with whom you want to communicate?
- *Where* do you want to reach them?
- *When* do you want to reach them?
- *What* information do you want to communicate to them?
- *How* many do you want to communicate with, *how* often, and *how* much do you want to spend to do it?

The answers to these questions establish your cable advertising objectives.

Next you need to establish a cable strategy by examining the degree to which cable can meet your objectives. The "Checklist of 'New Media' Media Values" (Table 5-2) provides a framework for this examination and strategy. It lists 18 media evaluation values to be considered and shows where these values are found in the "Non-New Media"—magazines, newspapers, out-of-home, radio, and television. Finally, it shows where these same values exist in the "New Media."

- Cable
    - Satellite Cable Networks
    - Regionally Interconnected Cable Systems
    - Individual Cable Systems
- Time-Shift Media
    - Video cassette Recorders
    - Video disc Players
- Information Systems (two-way Cable, Videotex)

TABLE 5-2 A Checklist of "New Media" Media Values

	Source of Values in "Non-New" Media					Opportunities in "New Media"				
						Cable				Info.
Media Evaluation Values	M	N	O	R	T	Sat.	Int.	Loc.	VCR Disc	Sys- tem
Who										
Audience Selectivity	X			X	X	X	X	X	X	X
Upscale Audience Profiles	X					X	X	X	X	X
What										
Visibility, Sound and Action					X	X	X	X	X	
Product Enhancing Environment	X				X	X		X	X	X
Product Demonstration Potential					X	X	X	X	X	
Flexible Ad Message Length	X	X				X	X	X	X	
Newsworthy Setting		X						X		X
Direct Response Generator	X	X			X	X	X	X		X
Quality Color Reproduction	X		X		X	X	X	X	X	
When										
Fast Dissemination of Information		X		X	X	X	X	X		
Long Life	X		X						X	
Short Notice for Late Buys		X		X	X	X	X			
Where										
Geographic Area Targeting	X	X	X	X	X		X	X		X
Highly Localized		X	X	X	X			X		X
Retail or Dealer Tie-In Potential		X			X	X	X	X	X	X
How (Many, Often, Much)										
High Reach Potential		X			X					
High Frequency of Message Delivery			X	X		X	X	X		
Low Unit Costs				X		X	X	X		

M - Magazine
N - Newspapers
O - Outdoor
R - Radio
T - Television
Sat. - Satellite
Int. - Interconnects
Loc. - Local

Finally there is the evaluation of the individual cable opportunities (the programs, the networks, and the cable systems) and then the actual buy and all of the subsequent follow-up. (See Figure 5-3.)

FIGURE 5-3 Developing Strategy for Selecting and Using the New Media

Identifiable Advertising Objectives	+	Identifiable Cable Opportunities	=	Intelligent New Media Usage

Implementing a Cable Advertising Program

In March 1981, the American Association of Advertising Agencies surveyed all offices of its member agencies to determine their procedures for buying cable television. At that time, one-third of those who responded indicated that their offices had been involved in a cable buy. Among those buying agencies, different individuals were responsible for making the buy. (See Table 5-3).

While the specific execution of a cable buy is a media function, the most effective cable advertising comes about through a very close working relationship among everyone involved in the buying and

TABLE 5-3 Who Buys Cable?

	Percent of	Breakdown by "Type" of Buy		
	All Buys	Local	Regional	Satellite
Spot Buyer	30	65	20	15
Network Buyer	19	—	6	94
Media Planner	37	43	6	51
Special Cable Buyer	6	20	10	70
Other	8	N/A	N/A	N/A
	100			

Source: AAAA, "Results of Survey on Cable Buying-Paying Procedures," August 1981.

selling of cable. This includes all those involved in evaluating it from a media standpoint, from a research standpoint, and from a creative standpoint. (See Figure 5-4, pg. 82).

Getting the Job Done

The setting and implementation of a sound cable advertising program follows an organized step-by-step procedure. Figure 5-5 and following figures show this process in a flow-chart format.

Step One

Examine your product or service in terms of its overall media and communications requirements. This will provide the basis for determining how well each cable opportunity can accomplish your objectives.

FIGURE 5-5 Setting Cable Advertising Strategy: Step One

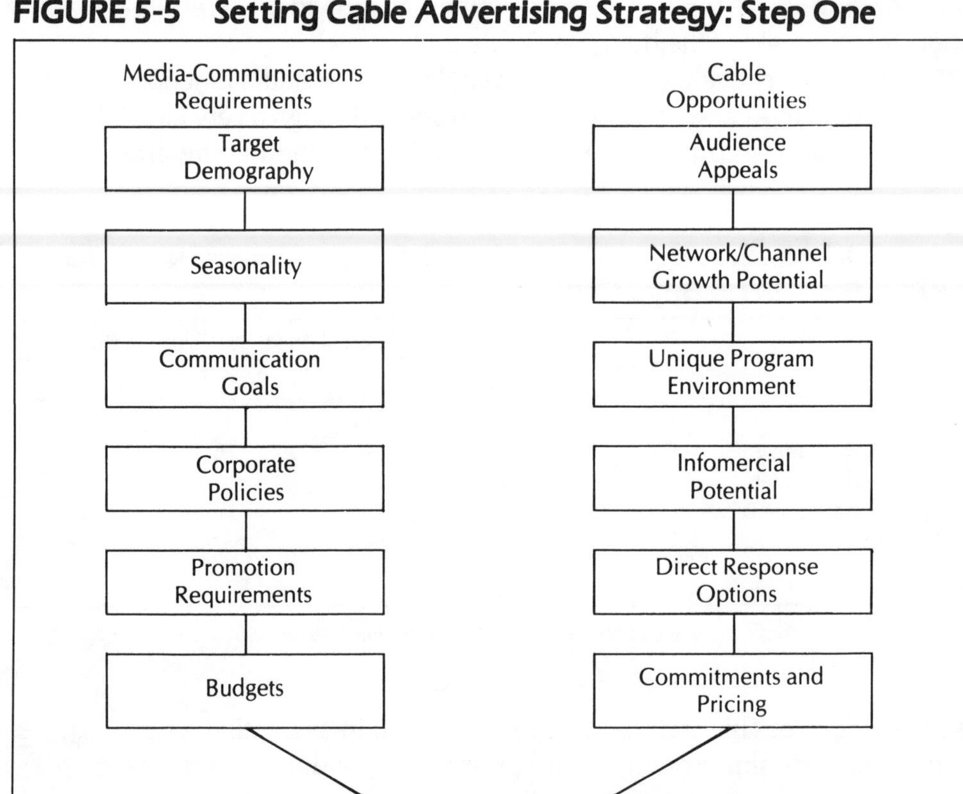

Step Two

In the negotiation stage, you will evaluate the specific cable proposal(s), negotiate the advertising schedule, and attempt to build in some form of long-term price protection to protect against sudden escalating costs. (See Figure 5-6.)

FIGURE 5-6 Implementing Cable Advertising Strategy: Step Two

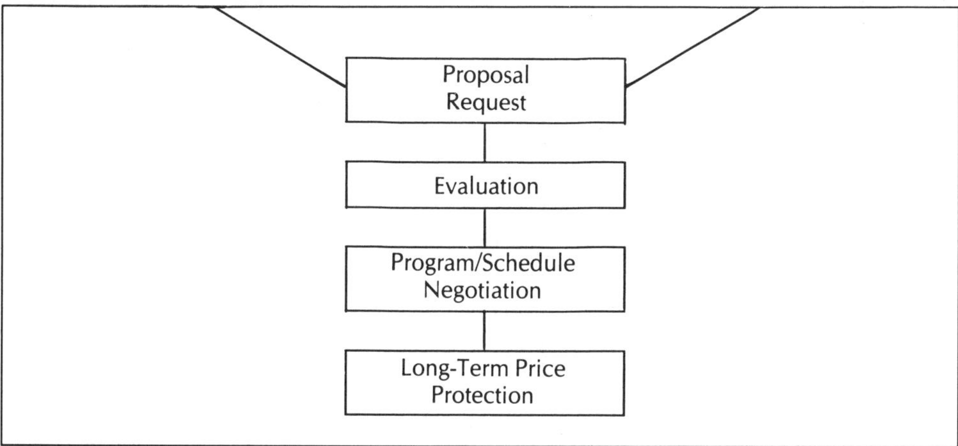

Step Three

The order is placed. Now it is time to prepare special commercials, plan publicity and promotion, and design any special research to measure the effectiveness of your cable schedule (Figure 5-7).

FIGURE 5-7 Implementing Cable Advertising Strategy: Step Three

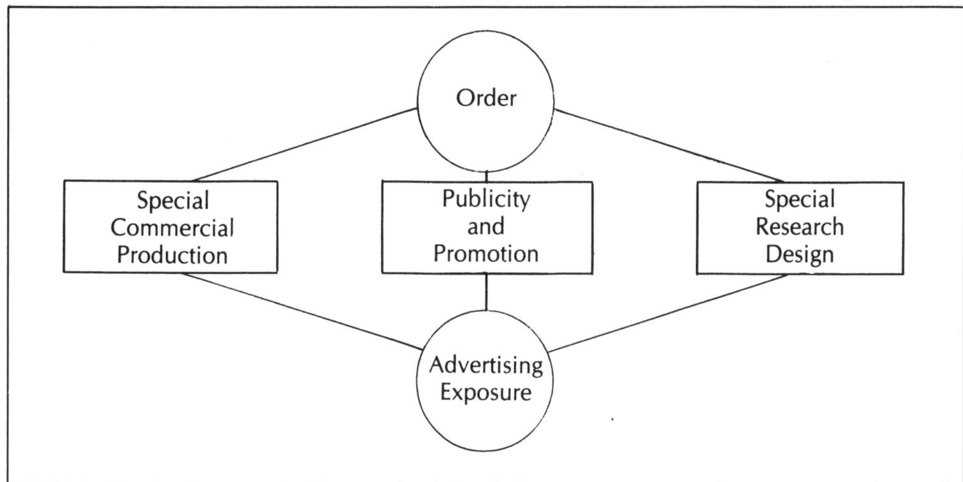

Step Four

Now, evaluate the response to your cable advertising either through audience research, direct response offers, or any special research you have designed. At the same time, you should continue to explore all other potential cable opportunities (Figure 5-8).

FIGURE 5-8 Implementing Cable Advertising Strategy: Step Four

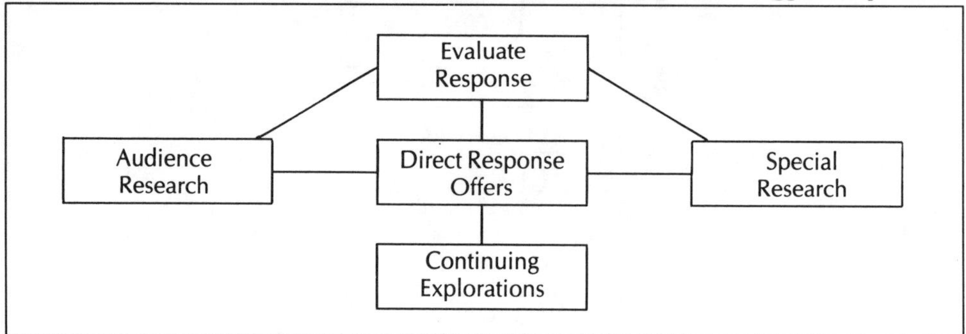

This planning strategy applies to local as well as to network cable. Advertisers can use it to direct their buying activity, and cable systems and networks with an understanding of it can do a better job of selling the medium. Figure 5-9 fully illustrates this process.

FIGURE 5-4 Making a Good Cable Buy

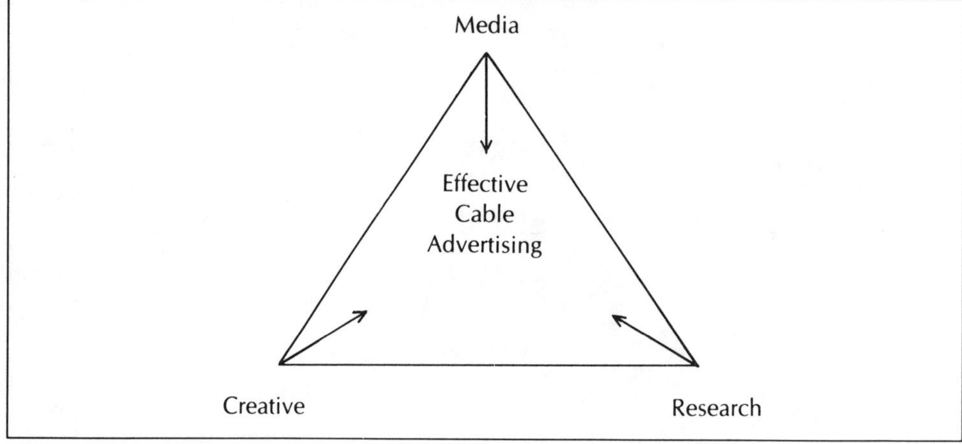

FIGURE 5-9 Setting and Implementing Cable Advertising Strategy

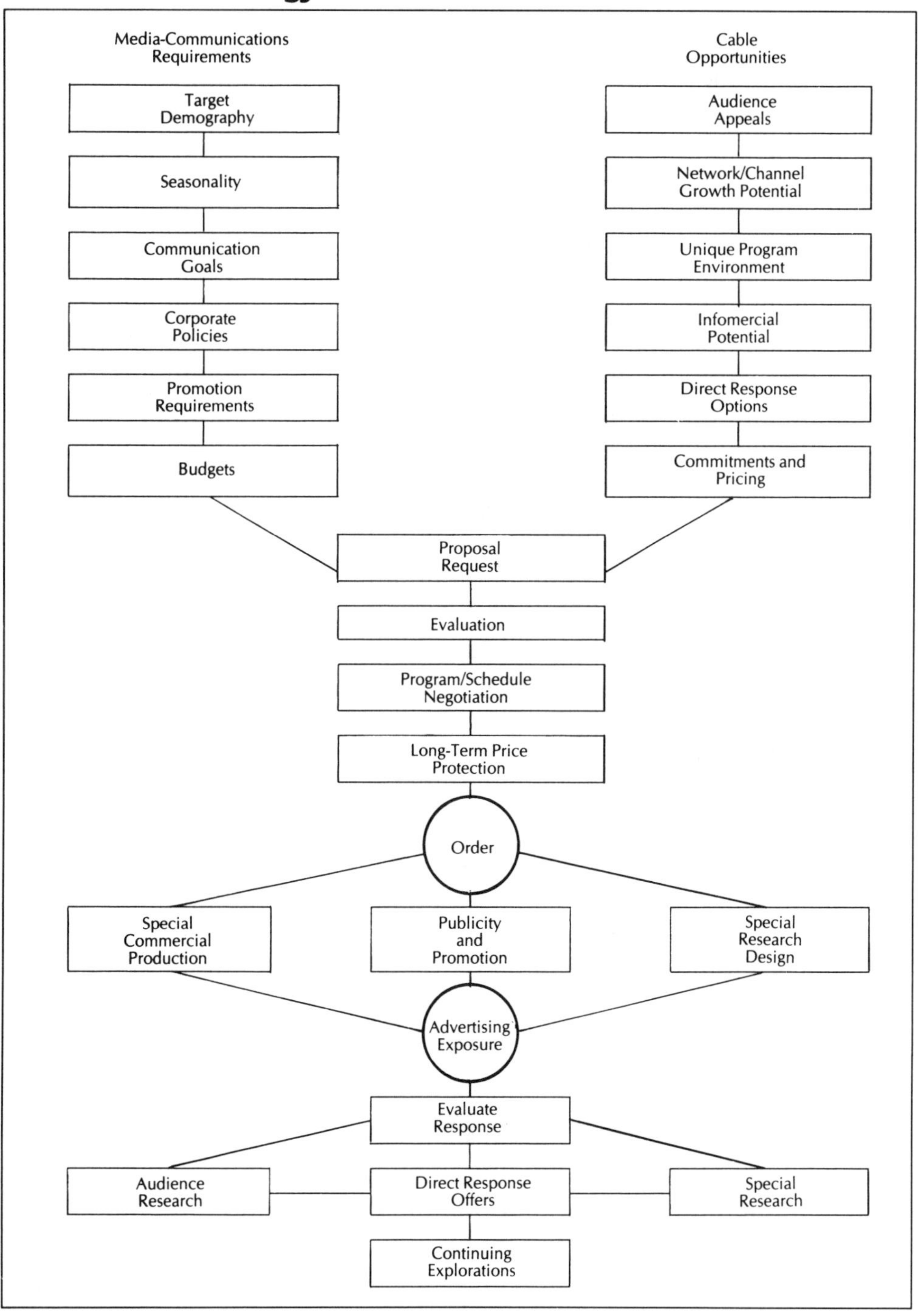

Achieving Cable's Maximum Benefits: Case Histories

Throughout the entire cable planning process, it is essential that an advertiser "keep on the right track." This involves a complete examination of what you want cable to accomplish, a thorough investigation of all cable opportunities available to do this, and a constant reexamination of what is being done.

In a nutshell, one must ask and re-ask:

1 What is our advertising media strategy accomplishing?

2 What would we like it to accomplish that it is not now doing?

3 What cable opportunities can help us do this?

4 Is our cable effort succeeding in doing this?

5 Are there additional cable opportunitites that should be considered?

The remainder of this chapter is devoted to brief case histories of companies that have followed the planning process described above. Although they are very different companies marketing very different products, they have at least two things in common. One, they identified their objective, thoroughly investigated ways of attaining that objective, and reexamined their strategy. Two, all were successful to the extent that they gained important new knowledge about a new medium at a price that will never be lower.

Advertisers have used cable to achieve any one or a combination of several communications benefits. Following are 15 case histories of how cable was used by different companies to deliver or secure:

- A specialized, targeted program environment.
- A flexible commercial in terms of length or form.
- Program sponsorships and product exclusivity.
- Direct response.
- A low-cost testing potential.

Table 5-4 shows how the advertisers in these cases accomplished these objectives.

TABLE 5-4 Achieving Cable's Maximum Benefits

	Anheuser-Busch Clydesdale Collection	Hallmark Kaleidoscope	Kawasaki and Music Television	Kemper Insurance Co. Golf	Ken-L Ration Pet Show	Ken-L Ration and Westminister	Kraft, Cookbooks and Recipes	McDonnell-Douglas DC-10	Old Spice Sports Quiz	Quaker Oats and CBS Cable	Raisinets and the Movies	Scott Value Line	3-in-1, Plastic Wood and I.D.s	20th Century Fox Promos	Wilton Cake Decorating
Specialized-Targeted Programming		X	X	X	X	X		X	X	X	X	X	X		X
Flexible Message Lengths and Form	X		X		X		X		X	X	X	X	X	X	X
Program Sponsorships		X		X	X	X		X		X					
Product Exclusivity		X		X	X	X				X		X			
Direct Response	X				X	X	X	X	X			X			X
Low Cost Testing Potential	X						X		X			X			X

Anheuser-Busch

Goal: To test the ability of television to sell its Clydesdale Collection, a mail order catalog line of etched crystal mugs and beer glasses, home entertaining gifts, and personal accessories. Previously, the Clydesdale Collection's advertising had been confined solely to magazines.

Cable Idea: A special 90-second cable advertising message was created for exposure on ESPN sports programming. Its aim was to increase product awareness, catalog requests, and sales. Cable permitted television to be tested at low media and commercial production costs.

Hallmark

Goal: To extend the reach and promotional impact of the "Kaleidoscope" show, which is staged at its headquarters in Kansas City and taken on tour to other cities. The show centered on crafts—how to make things and how to put things together.

Cable Idea: Hallmark produced a series of five-minute programs directly related to "Kaleidoscope." They aired them on the USA Network's "Calliope" to reach children across the country.

Kawasaki

Goal: To create a low-cost, unique commercial message targeted specifically at Kawasaki's young audience and closely associated with the style and feeling of the media vehicle for which it was designed.

Cable Idea: A two-minute "musical film," with an original music score written especially for MTV: Music Television, was created. Kawasaki motorcycles, of course, were featured. The production involved the re-editing of existing commercial elements—motorcycle film and stock footage—and the specially written rock music was based on the "Kawasaki Lets the Good Times Roll" theme.

Kemper Insurance Company

Goal: To provide additional advertising and promotional support for the annual Kemper Open Golf Tournament on the CBS Television Network.

Cable Idea: Each year, Kemper produces a special golf film based on the current year's tournament. The film is used by Kemper agents and made available to local clubs and organizations as well as to television stations.

Prior to the 1982 tournament, the 26-minute 1981 Kemper Open film was run three times on ESPN. It created a greater in-depth awareness of the event than could have been conveyed in shorter broadcast length tune-in announcements.

Ken-L Ration

Goal: To enhance Ken-L Ration's commercial message environment, strengthen its image among consumers, foster good will among veterinarians, and broaden distribution of its consumer promotional material.

Cable Idea: Ken-L Ration appear daily in the Cable News Network's "All About Pets," a 2-3 minute program segment covering topics on the care and feeding of pets.

A 15-second slide tag was added (at very little production cost) to Ken-L Ration's 30-second broadcast commercials. It offered Ken-L Ration's booklet on "How to Care for, Train, and Feed Your Dog" and reminded viewers about the importance of visiting their veterinarians regularly.

Ken-L Ration

Goal: To showcase Ken-L Ration in a unique environment related to dogs and targeted specifically toward dog owners.

Cable Idea: Sponsorship of the USA Network's six-hour coverage of the Westminister Kennel Club's Dog Show. The most prestigious and famous dog show in the country, it offered Ken-L Ration:

- A large concentration of dog owning viewers.
- A program environment geared 100 percent to the product.
- Exclusivity as the only dog food advertised in the event.
- Favorable publicity in the form of press releases in trade publications.

Kraft

Goal: To gain additional exposure for commercials whose previous airing had been limited to network television specials *and to* provide an opportunity for viewers to respond to write-in offers for Kraft recipes.

Cable Idea: The audio for "Baked Foods from Around the World," a 90-second Christmas holiday special, was retracked to offer a free recipe booklet by mail.

Following this, elements of several commercials were reassembled and re-tracked into a 90-second cable message offering Kraft's 75th Anniversary Cookbook for $4.95 to those viewers who wrote in for it. These commercials ran on the Cable News Network and WTBS.

McDonnell-Douglas

Goal: To boost the impact of its DC-10 advertising effort among *key* governmental and military decision makers as well as the public at large.

Cable Idea: On ESPN, McDonnell-Douglas sponsored the annual Army-Air Force football game. Its commercials were enhanced by the addition of a slide tag that offered a booklet with further information on the DC-10.

Old Spice

Goal: To create a compatible environment for this men's fragrance and determine (without existing rating data) which of several sports programs offered it the greatest audience attraction.

Cable Idea: Old Spice created a sports trivia contest that centered on events and stars in baseball, basketball, boxing, football, and hockey. The commercials ran in a variety of ESPN sports programming with each program's responses identified by a different code. Thus, Old Spice developed its own measure of audience appeal without waiting until rating data were available on ESPN.

The Quaker Oats Company

Goal: To demonstrate Quaker's commitment to cultural arts programming and reach an upscale target audience in new, interesting, and creative ways.

Cable Idea: Quaker purchased full sponsorship of CBS Cable's entire October 12, 1981, premier night of programming. From a creative standpoint, Quaker blended its existing food and pet commercials with specially created variable length messages that:

- Explained Quaker's dedication to quality, all-family programming.
- Described the care and expertise that went into the design and building of Fisher-Price toys.
- Told the story of a boy and his little lost dog.
- Presented the evolution of the Quaker logo.

Raisinets

Goal: To reach an under-30-year-old audience, which represents the primary consuming group, and to increase its association with movie going, where a large volume of the candy's consumption takes place.

Cable Idea: A schedule of 10-second and 30-second announcements were run on MTV: Music Television. The 30-second commercial was a humorous parody of movie scenes and a play on words in movie titles. It ran immediately adjacent to movie company spots on MTV: Music Television and created a complete movie environment for the product. The 10-second message featured the name *Raisinets* in a visual of a theater marquee sign. This commercial highlighted the fact that Raisinets are available in your local theater, and it then led into a theatrical movie commercial.

Scott Paper Company

Goal: To develop an advertising feature that could promote Scott products in an environment that demonstrated the company's service orientation to the consumer.

Cable Idea: Scott created the "Scott Value Center" for the Modern Satellite Network's "Home Shopping Show." Running twice a day, five days a week, these one-minute tips provided helpful information, promoted Scott's wide variety of products, and offered viewers booklets and product samples.

3-In-1 Oil and Plastic Wood (American Home Products)

Goal: To secure national exposure for a group of 10-second commercials that could only be aired as newsbreaks in network television or in local-market spot positions.

Cable Idea: The commercials were placed in male-oriented sports programming on ESPN and the USA networks. These cable networks had the flexibility to accept shorter length spots as well as long-form messages.

Twentieth Century-Fox

Goal: To promote the release of a new film, "Fort Apache, the Bronx," with more than the usual 30-second network television commercials.

Cable Idea: The full two-minute theater trailer with a 5-second tag ("Starts February 6 at theaters everywhere") was run on CNN, ESPN, and WTBS.

Wilton Enterprises

Goal: To provide a greater ability to demonstrate and explain Wilton's cake-decorating products and courses than was possible with its 30-second broadcast commercials.

Cable Idea: Wilton developed and aired a 9-minute segment for the "Home Shopping Show." It presented cake-decorating techniques and concluded with a beginner's book offer and a toll-free 800 number for futher information.

These 15 cases are just the beginning. Throughout this book, you'll find dozens more, including two special appendices with 66 examples of how advertisers have successfully used network cable and over 100 ideas for cable creativity at the local level.

CHAPTER 6

Different Avenues for Different Advertisers

The Scope of the Business: National or Local

All advertisers are not created equal. As a result, cable opportunities should be evaluated initially in relationship to the scope of each advertiser's business and communications objectives. Is the advertiser's business national or local? Is the advertiser's product or service of mass use and appeal, or is it more narrow? The answers to these questions provide the key to the successful use of cable as an advertising medium.

National Satellite Networks

Three avenues are open to the prospective cable advertiser. The advertiser-supported national satellite networks provide a variety of options, some that have relative mass appeal and others that are narrower in viewer selectivity. While the term *national satellite network* implies uniform coverage throughout the country, such is actually not the case. Until the larger metropolitan markets are wired, there still will be many soft spots. However, in a number of these areas, penetration is and will be increasing rapidly. And as was pointed out in a previous chapter, in many markets, cable penetration is low in the central cities but quite high in the surrounding suburbs.

The absence of high levels of cable coverage in certain large markets should not deter an advertiser from using the satellite-fed cable services *unless these markets represent primary areas of sales strength.* Despite limitations of coverage in some areas, the number and variety of advertiser-supported satellite program services are con-

TABLE 6-1 Advertiser-Supported Satellite Services: October 1984

Satellite Network	Programming	Subscribers(MM)
Arts & Entertainment	Performing arts plus BBC comedy, drama and specials	17.0
Black Entertainment Television(BET)	Black-oriented films, sports, music and women and youth features	7.5
Business Times	Early morning business and financial news updates and features	34.0
Cable News Network(CNN)	24-hour in-depth news and features	30.5
CNN Headline News	24-hour headline news service	14.5
C-SPAN	24-hour public affairs	18.5
CBN Cable Network	24-hour family entertainment	27.0
Country Music Television	24-hour country music videos in stereo	5.0
ESPN The Total Sports Network	24-hour sports	34.0
Financial News Network(FNN)	Weekday financial and business news, interviews and market reports	17.0
Lifetime	24-hour programming about health, relationships, and self development	20.5
MSN The Information Channel	Daily information and service features	10.5
MTV: Music Television	24-hour video music in stereo	24.5
Nashville Network	Country entertainment	19.5
Nickelodeon	Daily programs for young people	23.0
Satellite Program Network(SPN)	24-hour international, financial, lifestyle and how-to programming	12.0
USA Cable Network	24-hour sports, women's, children's and entertainment programming	27.0
The Weather Channel	24-hour live national, regional and local weather and weather features	15.5
WTBS Superstation	24-hour entertainment, movies and sports	33.0

tinually increasing. Table 6-1 lists 19 major advertiser-supported satellite services and their subscriber counts as of October 1984.

Interconnects

Cable interconnects represent several cable systems in a given area that join together to exchange cable programming and to sell advertising with the convenience of one-order, one-bill placement. They can be connected either electronically via a microwave hookup or by "bicycling" videotapes between participants.

Interconnects allow a national or local marketer to use cable to both add and replace other advertising weight going into an area. While an individual cable system might cover only a portion of a market, an interconnect may represent the equivalent of an entire ADI, DMA, or Standard Metropolitan Area. Interconnected systems have been referred to as CMOs—or Cable Markets of Opportunity—to distinguish them from broadcast television markets.

The extent of an interconnect's coverage area can stretch for a considerable distance. For example, the San Francisco Bay Area Interconnect reaches all the way from Napa in the North to Salinas in the South.

From the marketer's standpoint, an interconnect permits an advertising message to be targeted to the specific demographic or geographic audience he wants to reach. Commercials can automatically be slotted in the local advertising positions of cable satellite networks (CNN, ESPN, MTV, USA, The Weather Channel, etc.) and other programming carried by the individual cable systems in the area. The advertising can be transmitted to all of the interconnect's cable systems, or, in the case of the more highly developed interconnects, it can be directed only to those specific communities the advertiser wants to reach. In this case, a lawn care product might have its message directed only to systems in areas with a high incidence of single-family dwellings.

Looking ahead a few years, super-interconnects may be developed to serve larger regional areas, such as the entire West Coast. The technology is already available, but advertiser interest will determine how soon this is accomplished.

Somewhere between the national satellite networks and the regional interconnects are the regional pay sports networks, which are both advertiser and subscriber supported. Among them are Sports Time, Home Team Sports, New England Sports Network, SportsVue, and Cox Cable-San Diego Padre's pay-per-view service.

The most ambitious service launched in 1984 was Sports Time, which extended to 15 midwestern states and was backed by Anheuser-Busch, Multimedia, and Tele-Communications Inc.

Strictly Local Cable

Unlike broadcast television, where a station's economic survival depends on the advertising dollars it generates, cable's primary source of income comes from subscriber revenue. Thus, the majority of the nation's 6,000 cable systems have yet to take an aggressive stance in selling advertising, and the best estimates are that only between 15 percent and 20 percent of them are now doing so. This number, however, is rapidly rising as system owners have begun to see the potential profits that advertising can generate.

To date, most advertising support of cable has been with the national satellite networks. Its real, long-term potential, however, may be more at the local grassroots level. And, unlike broadcast television from Los Angeles, Omaha, or Atlanta, cable advertising will impact far differently on the cities and suburban communities it serves. It offers several distinct benefits to advertisers attempting to target local audiences:

- Since cable systems have carefully defined boundaries, the retail points affected by advertising can usually be identified. (For example, Orkin could not use television in New York since it covered areas outside of its service area. Cable, however, allowed Orkin to pinpoint specific communities that they did service.)
- An advertiser knows the area where advertising is being delivered, and there is no "spill-out" or "spill-in."
- The cable family is directly tied-in to the cable operator and receives at least one monthly mailing—a bill and/or a program schedule.
- With cable's many channels, there should be no shortage of time, as with network and spot television. As a result, an advertiser can expect to hold the line on costs and can expect greater flexibility in scheduling and in commercial design and length.

In a large metropolitan area, advertisers already have a full media menu to satisfy most of their needs. Chicago, for example, has 7

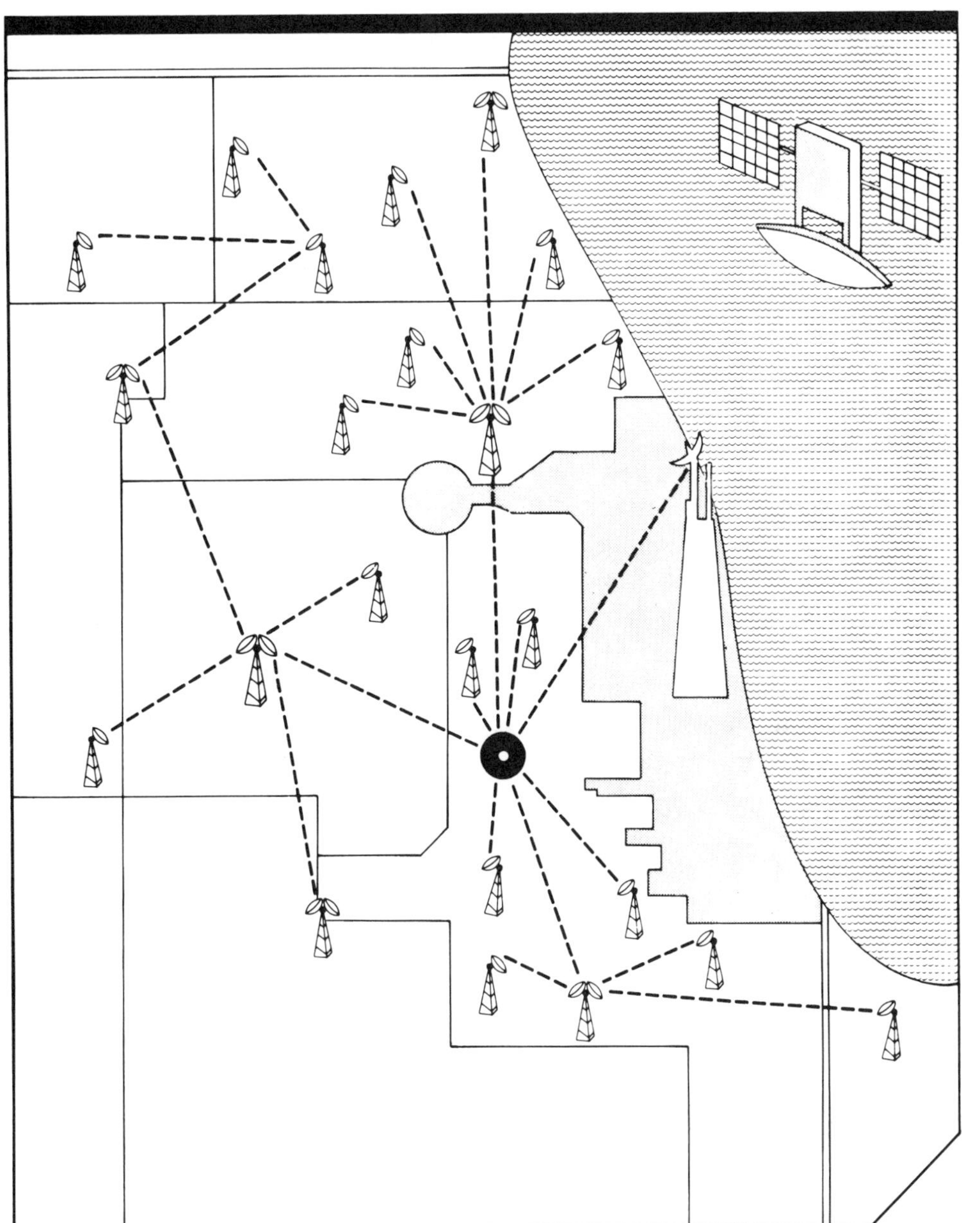

Chicago-area interconnect

commercial TV stations, 50 radio stations, a complete complement of general and ethnic newspapers, a city magazine, 4,300 billboards, posters, and car cards, and dozens of localized national magazine editions. Here, cable won't so much represent a totally new medium as it will a means of delivering a unique creative treatment that existing broadcast television does not provide for. This includes in-depth personal selling, advertising that is tailored to a specific program's environment, and affordable sponsorships as existed in the 1950s.

In the smaller towns and suburban communities, however, cable is a brand new communications form. In those areas, a franchisee, a dealer, or a retailer's media options have been largely limited to suburban newspapers, possibly a radio station, billboards, and direct mail. Television has not been practical either because its price tag or because coverage extended outside of an advertiser's area. Cable, however, offers an attractive alternative or addition to the local advertising media mix—low-cost, informational, service-oriented communication tailored to a specific group of prospects.

By 1990, nearly 60 percent of the nation's homes will be hooked up to cable. For the advertiser looking for the heaviest pockets of cable penetration today, Table 6-2 shows those Nielsen Coverage Areas that have already joined the 1990 "60 percent-Plus Club." In addition to these DMA's, there are many smaller communities and suburban areas with high levels of cable penetration and interest in attracting advertising revenue to their systems.

A new service of the A.C. Nielsen Co. is aimed specifically at providing advertisers, agencies, cable services, and multiple system operators with an easy way to access information on local cable subscribers and advertising availability. CODE (Cable Online Data Exchange) is able to retrieve 450 different kinds of informational areas which are constantly updated. In terms of advertising, CODE can provide the number of systems offering local ad availabilities and prevailing rates. It also will give the telephone number and address of the person in charge of advertising at each of the 776 cable systems. As for subscriber counts, CODE will provide the data by headend, system, tiers, county, state, cable network, geographical region, franchise and DMA.

The Nature of the Product or Service: Mass-Market or Non-Mass-Market

Most mass-marketed goods and services are relatively similar in use and in target audience. As a result, advertising has generally sought to differentiate them from a creative standpoint by developing distinct brand personalities and focusing attention on what Rosser Reeves referred to as the Unique Selling Proposition.

TABLE 6-2 Nielsen Coverage Areas with Over 60 percent Cable Penetration

San Angelo	86.0	Amarillo	65.7
Santa Barbara	79.1	Bakersfield	65.5
Laredo	77.8	Glendive	65.3
Marquette	77.0	San Diego	65.1
Parkersburg	75.7	Salisbury	65.0
Odessa-Midland	73.7	Greenwood-Greenville	64.7
Victoria	72.4	Honolulu	64.1
Beckley-Bluefield-Oak Hill	71.5	Binghamton	63.9
Johnstown-Altoona	71.4	Presque Isle	63.4
Yuma	71.3	Wichita Falls & Lawton	63.3
Ft. Myers-Naples	70.6	Wheeling-Steubenville	63.3
Biloxi	70.5	Wichita-Hutchinson	62.5
Utica	70.3	Reno	62.4
Monterey-Salinas	70.0	West Palm Beach	62.3
Eureka	69.9	Eugene	62.3
Clarksburg-Weston	69.6	Bend, Or	61.8
Roswell	69.3	Charleston-Huntington	61.8
Zanesville	68.8	Topeka	61.6
Lima	68.6	Champaign & Springfield-Decatur	61.0
Abilene-Sweetwater	68.6	Gainesville	60.5
Wilkes Barre-Scranton	68.2	Tyler	60.2
Casper-Riverton	67.0	Butte	60.1
Cheyenne-Scottsbluff	65.9	San Antonio	60.1

Source: A. C. Nielsen, November 1984

In terms of media selection, the differences in plans between competing products and services are often quite small. In radio and television, selection has largely been on the basis of age and sex. Only in print has there been a real effort to zero-in on highly selective audiences and environments through special interest magazines.

This relative "mass nature of the mass media" has often frustrated advertisers seeking to better concentrate their marketing dollars against selective marketing targets. And it has been even more frustrating to the smaller advertisers of non-mass-market products who, to paraphrase John Wanamaker, often feel that "at least half of their advertising dollars are wasted but aren't sure which half!"

Targeting Audiences and Messages

Cable video allows the mass-market advertiser to target more closely on specific marketing segments that he regards as having above-average consumption potential.

And for the small budget, non-mass-market advertisers, cable can provide a very precisely targeted video environment he might be unable to afford in network or spot television and has had to seek out in print. When this more highly targeted video environment is coupled with the use of a more informationally oriented, in-depth advertising message, the results can be an overall increase in advertising effectiveness. (See Figure 6-1.)

Following are 15 examples of the use of different cable avenues for different advertisers, recognizing that "all are not created equal" but have different objectives in terms of audience selectivity, program compatibility, and geographic targeting.

Lifetime Lifetime includes numerous programs delivering information on health and physical well-being. An insurance company could sponsor a show on nutrition, or a pharmaceutical company could sponsor an exercise show.

American Baby Cable Show The first cable program series devoted to educating the new and expectant mother, the American Baby Cable Show (SPN, CBN) provides information and live demonstrations of all phases of childbirth and early life. In addition to its cable exposure, the series is distributed free to childbirth classes throughout the country.

The Weather and Your Pet One example of how a cable network can work effectively with an advertiser and its advertising agency to develop a highly targeted and informative cable feature involves The Weather Channel, Ralston Purina, and Gardner Advertising. "Weatherize Your Pet" offers tips on protecting pets from inclement weather. It is an infomercial series that gives suggestions for preventive care during harsh winter weather conditions. Then as the seasons change, new appropriate messages are produced.

Family Guide to Boating Fun This newspaper supplement represented a new vehicle for boating advertising and combined manufacturers' national ads with local dealer tie-ins. It was coupled

with a point-of-purchase kit and contest drawing and a half-hour cable show (MSN) featuring products of the advertisers. It was a classic case of in-depth information directed to a concentrated (boat-owning) target audience.

The Hospital Satellite Network Hospital Satellite Network delivers original programming for both hospital medical staff and patients and offers teleconferencing capabilities to the medical community. Subscribing hospitals get education programs for the professional and medical staff, information and entertainment programs (movies and comedy features) for patients, and "access time" programming consisting of video conferences and special events.

Hospital Satellite Network is a subscriber-dependent with the service's income supplemented by advertising. Advertising can be

FIGURE 6-1 Media Communications Segments

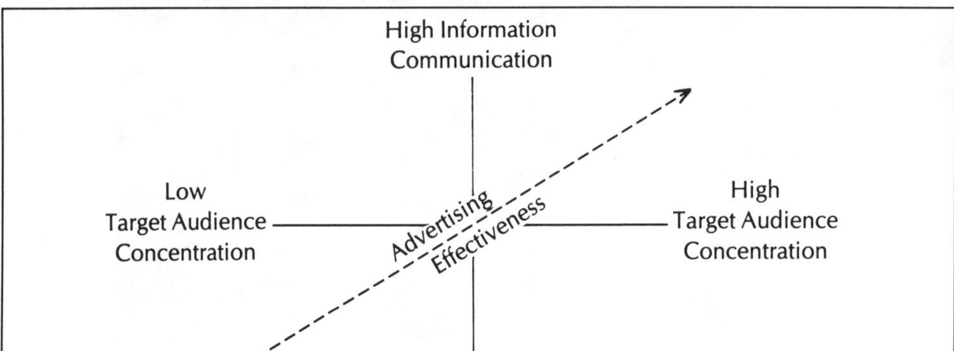

right on target for a wide variety of products and services aimed at a person's health and well-being as well as the needs of the hospital professional. In fact, an imaginative pharmaceutical company could even produce a series of infomercials on relaxation and exercise for the hospital patient. It could be accompanied by a booklet distributed to patients when they check in or with their meals.

Mitsubishi and the Mirage Bowl The 1981 Mirage Bowl (ESPN), which took place in Tokyo and featured Western Conference teams, San Diego State against the U.S. Air Force Academy, was sponsored by Mitsubishi. This is a good example of how Mitsubishi, a Japanese company, dominated a specific event on cable and provided advertising and promotion support for its products in a foreign market.

Military Cable TV Network The Military Cable TV Network is aimed at the active and retired military market and sells time on local cable systems that reach 54 military bases around the country. While the cable systems reach more than just the military audience (e.g., San Diego delivers a greater audience than the naval base alone), the concentration of military personnel is greater than would be offered by broadcast stations.

Cable, The Olympics, and Kentucky Fried Chicken The low-cost attraction of cable was a factor in Kentucky Fried Chicken's decision to advertise on 200 "Olympic Updates" and ESPN's coverage of the 1984 Winter and Summer Olympic Games. Kentucky Fried Chicken was able to secure an Olympic identity at a fraction of the cost of what Burger King and McDonald's paid for their ABC sponsorships, although of course, the audiences were substantially smaller.

Co-op and Cable for the Athletes Foot For many advertisers, cable has provided an effective and efficient means of using television nationally for the first time. The Athlete's Foot Chain of 475 athletic footwear stores, for example, used cable as its entry into the medium. Thirty-second spots featuring 15-second co-op inserts for shoe manufacturers appeared in ESPN's college and professional football games and the New York Marathon.

Sports Time Cable Network The importance of sports to the beer business was demonstrated by Anheuser-Busch's actual investment in Sports Time Cable Network, a regional pay service, along with partners Multimedia and Tele-Communications, Inc. Covering a 15 state, mid-America viewing area, Sports Time includes major league baseball, Big 10 and Big 8 College Basketball, regional NHL Hockey and NBA Basketball, professional Soccer, and other sports. Anheuser-Busch hopes to benefit from Sports Time Cable Network both as an investment and as an advertising vehicle.

H & R Block and Local Cable An H & R Block district manager conducted a Tax-A-Thon on his local cable system for 24 hours on the weekend immediately preceding the deadline for filing. In addition, he used the cable system in a variety of other ways:

- Tax tips were aired weekly on a community activities program.
- An annual H & R Block Banquet was taped and a 30-minute show was cablecast.
- Filmstrips of a new H & R Block office were shown.
- A live interview show was held.

- H & R Block executives were presented on cable.
- Tax-related items were carried on the weather scan.

"Woman's Day USA" A video translation of the monthly magazine, "Woman's Day USA" was produced by Young & Rubicam for its client General Foods and carried on the USA Network. Featuring a weekly menu planner, a shopping guide, and cooking features, it ran several times a week on Wednesdays through Saturdays (peak shopping days). The program did not, however, stand by itself. It was coupled with a number of two-minute "newsbreak" segments, "Today's Meal," that were scattered throughout the daily USA schedule. These provided instruction on how to prepare the key elements of the meals featured on the two-minute segments, and the two-minute segments enhanced the value

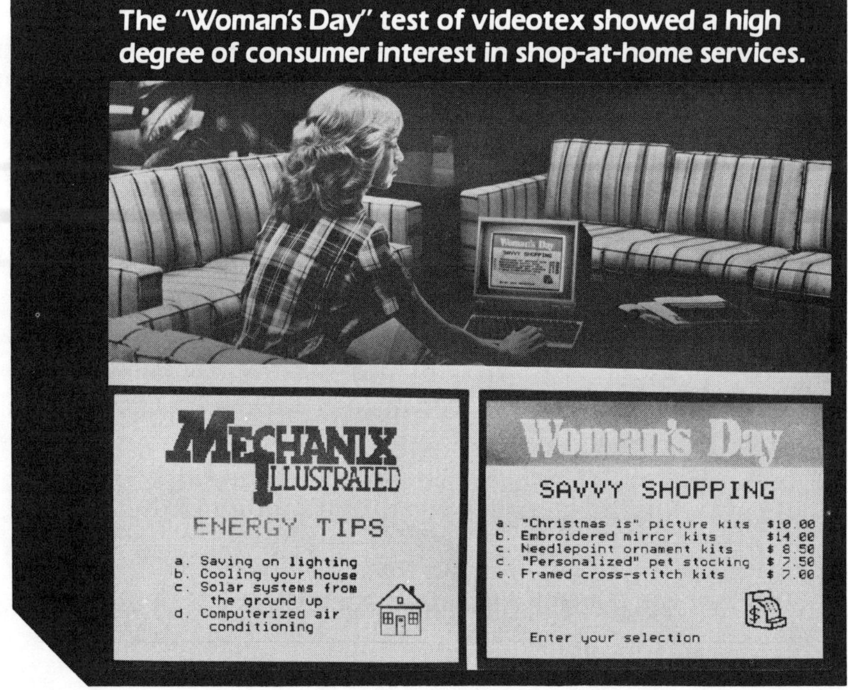

The "Woman's Day" test of videotex showed a high degree of consumer interest in shop-at-home services.

of the half hour. It is a fine example of new media synergy where one vehicle reinforces the value of another.

Rodale Press—Print and Video Synergy Rodale Press publishes a series of magazines aimed at helping people improve the quality of their food, health, homes, and lives. These include *Prevention, Organic Gardening, Bicycling, Rodale's New Shelter,* and *The New Farm.* Rodale brought the subject matter of these publications to cable in Rodale's "Home Dynamics," a 13-part series on the CBN Satellite Network. It focused on how to produce your own low-cost food and energy at home, thus allowing advertisers to reach this audience both via magazines and video.

To provide an even further extension of this multimedia approach to advertising, most of the programs in the "Home Dynamics" series are available as home video cassettes. Sponsors of the cable series have the opportunity of placing their advertising in these cassettes. These are then offered to groups like gardening clubs, solar energy societies, retirees' associations, etc.

Quaker Oats and the Weather Channel The Weather Channel provides national, regional, and local forecasts, along with weather-oriented features.

As a charter national advertiser on The Weather Channel, Quaker Oats sponsors special "cold weather reports" to promote its hot cereal products when the temperature dips below 40 degrees in a large number of markets. In the past, advertisers have developed weather associated features to run on radio when the temperature hit a certain level, when it rained, or when snow was being forecast. The Weather Channel allows this audio advertising to become video oriented.

The Sears Cable Catalog In a unique approach to merchandising its catalogs, Sears presented "Pretty Pictures (The Wishbook on Location)" as a Modern Satellite Network Special in summer 1983. The 30-minute show provided viewers with an insider's look at the world of fashion photography and the creation of photographs to be used for the Sears catalog. It showed how, working from an artist's layout, the photographer and model combine attractive fashions, beautiful locations, and a few tricks of the trade to make the book something special.

Two Approaches to Cable: Media Audience and Response

The Media Audience Approach

Media are traditionally evaluated on the basis of the number of potential prospects for a product or service an advertiser can reasonably expect to be in the audience of the programs or publications where the advertising appears. For television this generally means talking about the millions of impressions or the reach and frequency delivered against viewers of various demographic characteristics. From a pure media standpoint, cable can provide advertisers with low-cost continuity of exposure they may not be able to afford on broadcast television. An advertiser might find that the elimination of only one commercial in a television schedule will allow him to purchase an extended cable flight aimed at his target audience. For example, for the $150,000 it cost for one network commercial on ABC's Monday Night Football, an advertiser could have purchased an 8-week 1984 schedule of 10 spots per week on ESPN, the 24-hour-a-day sports network.

Another interesting comparison is what might happen if an advertiser with two 30-second commercials (at $500,000 each) in the 1985 Super Bowl converted one of these to a schedule on ESPN.

- With two 30's in the Super Bowl, he would reach 60 percent of all television homes an average of 1.7 times each
- With one commercial in the Super Bowl and a $500,000 schedule on ESPN, his reach would remain virtually unchanged at 58 percent, while the average frequency would increase to 2.8 times.
- Most important, among the upscale pay cable households, his reach would rise from 62 percent to 69 percent and the average frequency of exposure would climb 255 percent from 1.8 to 4.6 times.

In a nutshell, he maintains his merchandisable exposure in the Super Bowl, but adds to it heightened efficiency and frequency of contact.

All evidence points to the continued growth of cable and the increasing channel capacity that will provide many viewers with 50 or more options to choose from. (Table 6-3 gives a 52-channel cable prototype.) And this increasing channel capacity, coupled with the expansion of cable and other over-the-air new media into new markets, will result in a year-to-year decline in broadcast audience ratings. For example, the situation with prime-time network television might look something like that shown in Table 6-4.

TABLE 6-3 A 52 Channel Cable Prototype

3-Local Net Affiliates	3-Ethnic
3-Indies	3-Religious
2-News	2-Children's
2-Sports	3-Education and Information
4-Imported Stations	4-Local Access
4-Movies	2-Music
2-PBS	1-"How To"
4-Entertainment and Culture	1-Women's
1-Weather	1-Games and Participation
1-Health	1-Info Retrieval
1-Senior Citizens	1-Business and Finance
2-Shopping and Infomercial	1-Country

As would be expected, the impact of the new media will be the greatest in those homes that have the most of it. Today, this means in pay cable homes. It also will be greatest during the warm weather network rerun season (see Table 6-5). And, finally, across the day there are significant differences, with network ratings suffering the most in pay cable homes in prime-time and early fringe hours (Table 6-6).

The full impact of cable on broadcast audience levels is even more evident when one examines individual markets with different levels of cable penetration. The result is that an advertiser buying a television schedule, whether nationally or locally, can expect to find substantially lower ratings delivered in cable households than in noncable households. While the differences will vary by market and by time of day, Table 6-7 shows what the situation might be in a time and in a place where 40 percent of the homes have cable.

The differences in Table 6-7 are seen not only in the ratings but in the reach and frequency of a specific broadcast schedule. For example, if an advertiser in the past looked only at the delivery of his plan among all homes in a market or in the country (including both

TABLE 6-4 A 10-Year Forecast of New Media Impact on Network Audiences (Percent)

		Primetime Network TV Audience Estimates		
	Cable Penetration	Homes Using Video	3-Network Shares	Average Network Rating
1981	28	59	83	16.3
1982	35	59	78	15.4
1983	41	60	77	15.4
1984	45	60	76	15.2
1985	48	60	73	14.6
1986	51	61	70	14.2
1987	53	61	68	13.8
1988	55	61	66	13.4
1989	57	62	63	13.0
1990	58	62	60	12.4

cable and noncable), the numbers might look like those in Table 6-8. Among cable homes, however, the total ratings, the reach, and the frequency all fall below the level in Table 6-8. (See Table 6-9.)

To compensate for this underdelivery in cable homes, an advertiser can buy spots either locally or nationally on the cable satellite networks. Ted Bates & Co. was the first national advertising agency to recommend to its clients that 5 percent (later increased to 7 percent) of their television network prime-time budgets be transferred to Superstation WTBS as a way of recouping audiences that the networks had lost to pay cable homes. The recommendation followed the analysis by Bates of a number of special Nielsen studies that

TABLE 6-5 1983 Prime-Time Network Ratings in Cable and Noncable Homes (Percent)

	Winter (February)	Spring (May)	Summer (July)	Fall (November)
Noncable Homes	19.1	15.8	13.3	17.7
Basic Cable Homes	17.3	13.7	10.8	15.6
Pay Cable Homes	16.5	13.3	9.8	15.8
All Homes	18.3	14.9	12.1	16.9

Source: Nielsen, *Cable TV Status Report*

showed a significant underdelivery of network schedules in pay cable households.

An example of how this theory would work in practice is shown in Table 6-10. A three-network television schedule of 34 day-time and 10 prime-time commercials delivered audience levels that were lower in pay cable homes than in all television homes.

The network television schedule was reduced by three daytime commercials, and in their place were purchased (for the same dollars) 23 commercials on WTBS. As a result of the addition of WTBS to the schedule, audience delivery among all television homes remained virtually unchanged. Delivery of pay households, however, increased so it was now on par with that of all households.

TABLE 6-6 1983 Cable and Noncable Home Network Ratings for Different Dayparts

	Primetime (Percent)	Daytime (Percent)	Early Fringe (Percent)	Late Fringe (Percent)
Noncable Homes	16.5	6.3	11.6	5.4
Basic Cable Homes	14.4	5.9	13.9	4.5
Pay Cable Homes	13.9	5.7	9.4	4.7
All Homes	15.6	6.1	11.1	5.1

Source: Nielsen, *Cable TV Status Report* (February, May, July, November)

WTBS was specifically recommended because it offered the largest cable audience on a national basis. Advertisers interested in reaching more specialized cable audiences, however, might also use other channels, of a more special interest nature to help fill this defi-

TABLE 6-7 Possible Ratings for Cable and Noncable Homes (Percent)

	All TV Homes (100%)	Noncable Homes (60%)	Cable Homes (40%)
Primetime Homes Using TV	60	59	62
Local Shares	85	95	70
Local Ratings Available	51	56	43
Average Network Rating	17	19	14

Source: Arbitron Special Report, "The Impact of Cable on Spot Television Buying," 1981.

ciency. Or, for the broadest possible reach, multiple cable network combinations can be put together.

ESPN, the 24 hour-a-day cable sports service, has demonstrated what happens when a network prime-time and sports schedule is cut back 15 percent with ESPN substituted in place of it. Among all men, there were 13 percent more GRPs delivered. In pay cable homes there were 70 percent more men GRPs delivered, and the percentage of men reached three or more times increased from 8 percent to 20 percent. (See Table 6-11.)

TABLE 6-8 Past Four-Week Reach and Frequency

	All TV Homes (100%)
GRPs a Week	120
Reach	80%
Average Frequency	6.0
Total GRPs	480

Source: Arbitron Special Report, 1981.

The Response Approach

Many marketers recognize the importance of advertising response rather than simple audience contact. They do not necessarily evaluate a media buy on the basis of the cost per thousand people it reaches. Rather, they are more concerned with the cost per response generated.

An example of the use of this approach could involve a local auto dealer who has an opportunity to do a 15-minute cable interview for $200 on the subject of "How to Buy a New Car." If 10,000 homes subscribed to the cable system and only 50 people watched

TABLE 6-9 Present Four-Week Reach and Frequency

	All TV Homes (100%)	Noncable Homes (60%)	Cable Homes (40%)
GRPs a Week	120	130	97
Reach	80%	84%	72%
Average Frequency	6.0	6.2	5.4
Total GRPs	480	520	386

Source: Arbitron Special Report, 1981.

TABLE 6-10 The Ted Bates Concept

	Traditional 3-Network Schedule	
	All Homes	Pay Cable Homes
GRPs	293 (100)	270 (92)
Reach	75% (100)	73% (97)
Average Frequency	3.9 (100)	3.7 (95)
Effective (3+) Reach	39% (100)	36% (92)
	3-Network Plus WTBS Schedule	
GRPs	295 (100)	296 (100)
Reach	76% (100)	76% (100)
Average Frequency	3.9 (100)	3.9 (100)
Effective (3+) Reach	40% (100)	40% (100)

Source: Special Nielsen Tabulation and WTBS

the auto dealer, most media planners would consider the buy both ineffective and inefficient: ineffective because it only reached 50 people out of 10,000 homes (less than a 1 percent rating), inefficient because, at a cost of $200, its cost per thousand was a whopping $4,000!

The astute auto dealer would, however, have looked at this as an outstanding advertising response generator. First, he got his message across to a full 50 people, far more than came into the showroom on a single day. And, second, it only cost $4 to talk to each of these people for 15 minutes. He would probably have been willing to pay two or three times this amount to anyone who came into his showroom and listened to him discuss how to buy a new car for a quarter of an hour.

In another case, a pet food manufacturer might seek a unique environment in which to tell the story about his new, premium-priced dog food. He has the opportunity to sponsor a prestigious dog show on cable, but on examining the potential audience delivery, he finds that it will cost at least $20 per thousand women reached—four times that of his current television plan. Is he discouraged? Not necessarily. First, he knows that the show provides a far more compatible environment for the product than his normal network of spot television programming. And, second, he feels confident that he is reaching a responsive audience and that the majority of the viewers will own a dog and represent good prospects for a premium dog food. As a matter of fact, if he were to zero-in on his basic late-fringe television buy and go beyond CPM women to CPM dog-owning women in

TABLE 6-11 The Impact of ESPN in Reaching Men

	All Adult Men		
	GRPs	Reach	Effective Reach*
Original Network Schedule	106	60%	12%
Reduced Network + ESPN	120	60	17
% Difference	+13%	0	+42%
	Adult Men in Pay Cable Homes		
	GRPs	Reach	Effective Reach*
Original Network Schedule	88	55%	8%
Reduced Network + ESPN	150	65	20
% Difference	+70%	+18%	+150%

*Three or more times

Source: Special Nielsen Tabulation (November 1982) and ESPN Estimates

an age group and income group that might be regarded as true prospects, his dog show cable opportunity looks like a real bargain. (See Figure 6-2.)

Cable and the Direct Marketer

For years it has been recognized that no one knows how to motivate sales more or measure the impact of advertising on sales more than does the direct marketer. The national advertiser, for example, generally must evaluate media effectiveness on the basis of the size of the audience reached relative to cost per thousand, consumer awareness studies, and product share trends. The direct marketer, however, can

FIGURE 6-2 A Late-Fringe Television Buy vs. a Dog Show

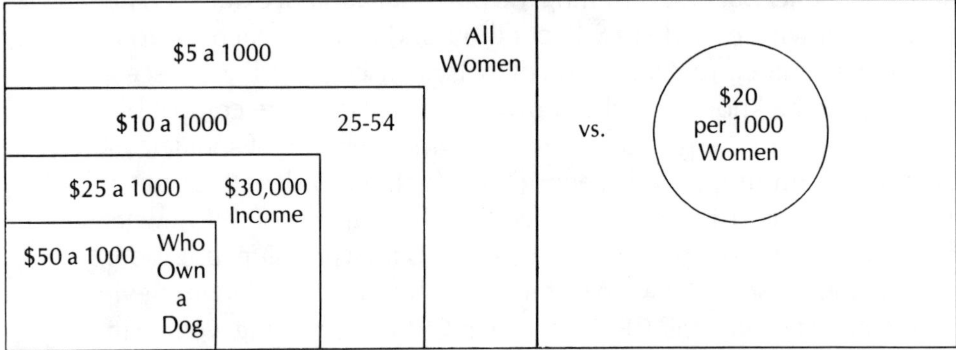

bypass the normal media measurement services and evaluate the effectiveness of individual vehicles on the basis of the sales or sales leads they produce. Over a period of time the direct marketer sees what works and what doesn't work and eliminates the ineffective vehicles from the media mix. (It has always been a bit amusing that direct marketers generally regard the television time periods that national advertisers feel are least attractive as extremely profitable in terms of pay-out per dollar expended—namely, late night and Sunday morning.)

Direct marketing has grown at a very healthy pace over the years, with sales volume up from $60 billion in 1975 to $150 billion in 1983. (See Table 6-12.)

TABLE 6-12 Estimated total U.S. direct marketing sales volume 1975-83

	($billions)
1975	$ 60
1976	$ 75
1977	$ 82
1978	$ 87
1979	$ 99
1980	$112
1981	$125
1982	$138
1983	$150

Source: Direct Marketing Assn.

The direct marketer uses a wide variety of selling tools, with 1983 advertising expenditures totaling over $30 billion for couponing, direct mail, magazines, newspapers, television, and the telephone. And even more is on the horizon, with experiments today with shopping via interactive cable television, videotex, teletext, and home computers. (See 6-13.)

The targeted appeals of the new electronic media coupled with their extended message length flexibility have made them extremely attractive to the direct marketing industry. And just as the direct mail order business is growing at a rate nearly twice that of retail, it would not be surprising to see its use of cable increase equally fast.

TABLE 6-13 Estimated total direct advertising expenditures (1977-83)

	(in $ millions)						
	1983	1982	1981	1980	1979	1978	1977
Coupons	182.1	127.1	94.6	84.2	72.0	61.0	84.0
Direct mail	12,692.2	11,359.4	10,566.7	9,998.7	8,876.7	7,298.2	6,966.7
Consumer magazines	188.7	167.0	150.0	135.0	123.0	99.8	86.2
Business magazines	73.9	66.0	59.0	53.0	47.0	49.4	49.4
Newspapers	80.5	70.6	73.0	60.6	54.4	58.0	42.8
Newspaper preprints	2,850.0	2,500.0	2,288.5	2,032.4	1,779.5	1,390.0	1,086.0
Telephone	13,608.3	12,935.6	11,467.0	9,845.0	8,555.6	8,555.6	7,699.0
Television	386.5	339.0	295.0	253.0	217.0	265.0	340.7
Radio	37.0	33.0	29.0	26.0	23.0	N/A	N/A
Total	30,099.2	27,597.7	25,022.8	22,487.9	19,748.2	17,777.0	16,354.8

Source: Direct Marketing Assn. Note: Creative costs not included in any of the above figures.

The Appeal of Cable to Direct Marketers

The segmented appeals of the special interest cable networks can be an effective direct line of sales for direct mail marketers. Perhaps one day there will even be "direct video lists," just as there are direct mail lists today. Fanciful? Mail order financial and insurance companies already can use business news channels to get new clients. Direct-marketed kitchen and houseware products find receptive audiences among the how-to and daytime women's service shows, and physical fitness products can utilize the cable health and medical networks.

The 800 telephone numbers, which today allow viewers to purchase the products they see on television or cable, may be the predecessor of the pushbutton interactive technology that will be developing more fully in the second half of this decade.

Direct response marketers will find the segmented appeals of cable coupled with the attractive ability to use nonstandard commercial lengths. In short, they will have the time required to convey the information they need to convey.

Direct marketing can be tied into cable in a number of ways to heighten its impact and extend its effectiveness. David Soskin, President of Soskin/Thompson Associates, a direct marketing advertising firm, pinpointed eight ways to maximize its effectiveness.

- Determine if cable subscriber mailing lists are available to support your cable advertising efforts.

- If your objective is to generate retail traffic or qualified sales leads, localize, humanize, and personalize your advertising message whenever possible.
- Test different cable dayparts, programs and seasonal mixes relative to broadcast television, print, and direct mail patterns. Cable response is often very different.
- Test and re-test, but always concentrate on the four major factors of message length, offer, copy, and graphics.
- To maximize response, consider using 800 Single Number Service to eliminate the need for a separate number for in-state callers.
- Ask for the order. Offer check and credit card options. Be clear and concise. Repeat the offer and how to get it at least twice.
- If the product or service has to be touched, felt, smelled, or explained in great depth, use direct mail or a telephone follow-up as an "interactive sale closer."
- Integrate your cable, local broadcast, print and direct mail campaign in a strategically and creatively consistent manner.

"The Great Catalogue Guide": A Unique Cable Test

In late 1980, the Direct Mail Marketing Association sought an innovative and dramatic communications program directed to consumers to achieve the following objectives:

1. Communicate the benefits of shopping by mail, e.g. convenience, value, selection, etc.
2. Devise a convenient and interesting way for consumers to identify and contact individual mail order catalog firms offering the products they want.
3. Generate 2,000 requests for this information.
4. Test cable's ability to achieve this within a reasonable budget.

A 120-second direct response commercial offered, through an 800 toll-free number, the DMMA "Great Catalogue Guide" as a free premium. The "Great Catalogue Guide" listed more than 550 consumer catalogs in 27 product categories. Postcards for consumers to fill out to request catalogs they would like to receive were also included in the "Guide." The commercial was transmitted by satellite to 600 local cable systems, which aired it from October 20 to November 3, 1980. The total budget allocated to the program (the commercial, the "Catalogue Guide," fulfillment, and media) was approximately $22,000.

More than 7,000 consumers called during the two-week period the commercial was aired to request the "Catalogue Guide," a 250 percent increase over the goal of 2,000 requests. This was a cost per inquiry of about $3. In addition to the excellent response, reaction from DMMA membership was overwhelmingly positive and supportive.

One Marketer's Opinion

Alvin Eicoff, President of A. Eicoff and Co. (now a part of Ogilvy & Mather), summed up the value of cable by saying:

> Broadcasting's 30- and 60-second commercials are artificial. Commercials should be able to be as long as they need to be and, as cable outlets develop, significantly longer commercial spots will become available. Many ad execs see such developments working to make it possible to sell such big-ticket items as automobiles via televised direct response ads. You'll be able to take the time to show the viewer the car's engine, explain why rack and pinion steering is the best, and tell the viewer what is so good about independent suspension. That would be a lot better than just showing a big cat jumping on a car! (*Advertising Age,* November 16, 1981, p. S-14.)

Cable and Direct Mail in Tandem

In combination, cable and direct mail can be an extremely powerful media package. For example, a 12-part cooking course can be created for cable at a very modest cost, perhaps even using the home economics department at a university. The course could be directly tied-in with a direct mail program offering the recipes, product coupons, a cookbook, and even "Videocooking Tapes." On a local basis, this could even be tied in with special grocery store promotions. The concept need not be confined to cooking but could be extended to how-to courses on car care, on fashion and grooming, or on health maintenance.

Sports Shack: A Success Story

One of the most successful direct response cable campaigns involved the Martin Lambert Advertising Agency and its client, Sports Shack. Sports Shack was a company involved in selling sporting goods store franchises. Their advertising was confined largely to newspapers and business publications. The company avoided using television because of its absolute cost and because they felt that the number of persons interested in and able to afford a $36,000 franchise was minute.

Sports Shack agreed, however, to test a flight of 15 60-second spots on the Entertainment and Sports Programming Network. If as many as 30 to 45 leads had been secured (2-3 per commercial), the campaign would have been regarded as a success. By the time the flight ended, over 600 people had called for further information. Of these, 86 requested interviews with the Sports Shack field sales staff. And, of these, 40 prospects agreed to fly to Minneapolis at their own expense to finalize contracts. For Sports Shack, cable proved to be an excellent response medium.

Marketing Politics by Cable

Politics has often been referred as the ultimate in direct marketing. In 1984, political strategists discovered that cable could be an important medium to market their candidates, and cable programmers and operators recognized that political advertisements offered them an important new source of revenue.

In their study, "Political Advertising and the New Electronic Media," Paley Communications, Inc., noted that several factors were contributing to the growing importance of political advertising for cable television. For example, the targetcasting of different programming services, whose audiences are more segmented than are those of broadcast television meant that a candidate could place his ads on several programming channels and target each message to a distinct group of "viewer-voters." Political candidates also can use long-form infomercials and utilize aspects of cable's interactive capabilities. *And* of special importance also during the high TV demand Election-Olympic years is the ability to take advantage of cable's low cost and high availability.

Information in the Home

It will be a long time before it is determined just how successful the new media can be in encouraging people to shop from their living room chairs. Some people will like the idea, others won't. And, for some products, it will be successful, while for others it will probably be disastrous. Aside from actual product purchases, however, the television set can be used to distribute many forms of consumer information.

Advertising, Videotex, and Teletext

For advertisers interested in providing consumer information and transaction services, videotex and teletext offer a number of opportunities. Experiments already have been conducted by many companies, including CBS, AT&T, Knight-Ridder Newspapers, Dow-Jones and Co., Reader's Digest, Cox Cable, and the Times-Mirror Company.

In Ridgewood, New Jersey, CBS and AT&T joined together in the fall of 1982 to test for seven months the ability of videotex to achieve consumer acceptance in a number of different areas including:

- To call up information such as news, sports, weather, financial data, encyclopedias, almanacs, schedules, etc.
- To conduct transactions such as banking, bill paying, shopping, making travel and entertainment reservations.
- To participate in computer-based education.
- To be entertained by computer-based games, quizzes, and features.
- To communicate electronically with other users—a technique known as electronic mail.

This is but one of a number of experiments with computer-based home information systems that have been and are going on around the country.

After nearly two years of development and testing, Time Inc. closed down its cable-based teletext project in November 1983. A test of Time Teletext was conducted for a brief period in Orlando,

Keyfax Interactive Information Service

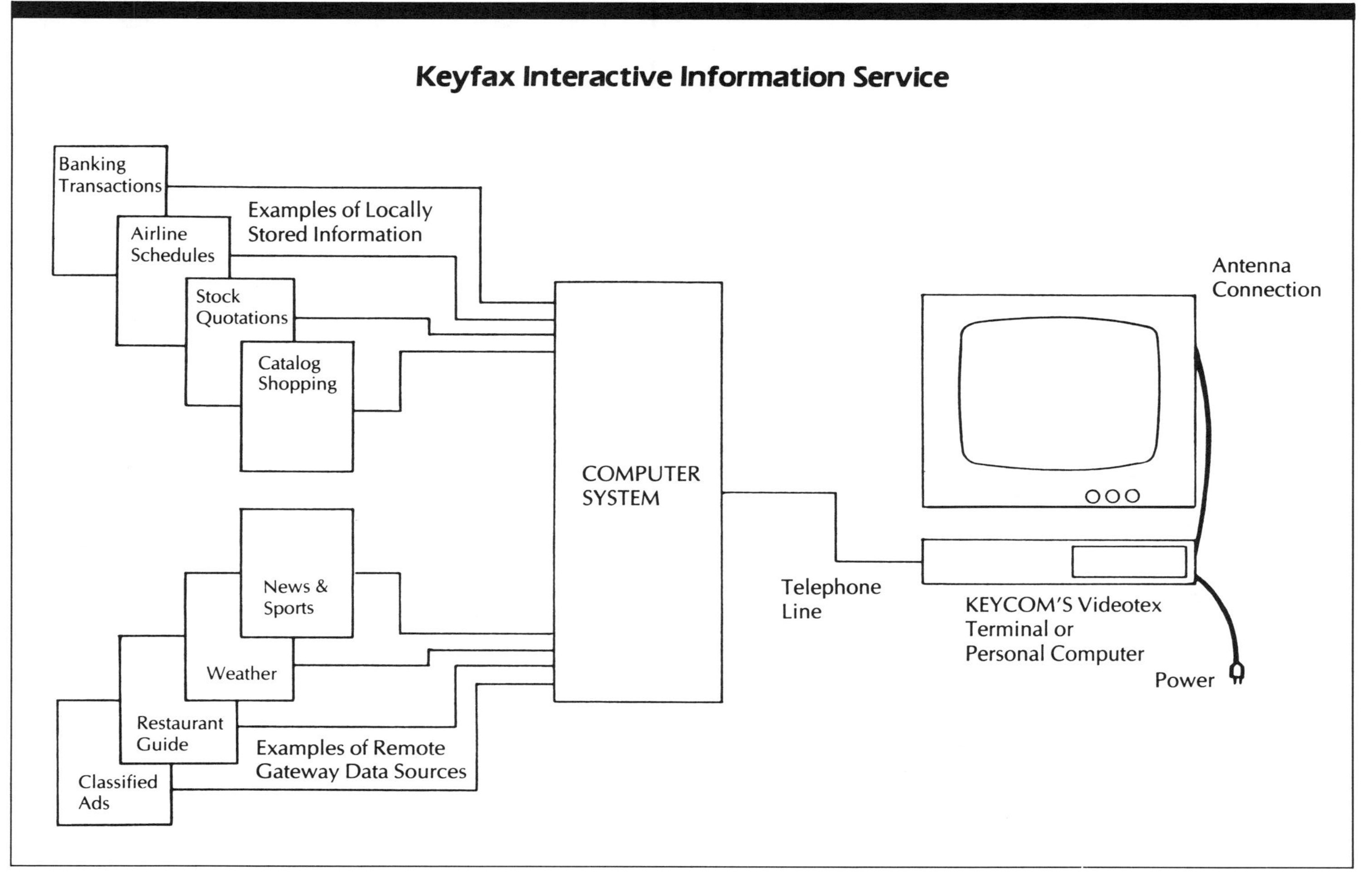

Florida, and San Diego, California, involving the participation of a number of major national advertisers. Time's decision did not mean that the company was totally giving up entry into electronic publishing. Rather, it reflected that the economics of the system were such that it could not be developed at reasonable prices in a reasonable period of time.

Although Time decided to halt its teletext plans, other media companies are moving ahead with their projects. CBS and NBC have national projects in the works, in conjunction with their affiliates, and Taft Broadcasting is continuing with its Cincinnati teletext experiment. In the Chicago area, the Keyfax Interactive Information Service is being tested. In addition, Times Mirror Co. has a commercial teletext venture named Gateway under way in southern California. And Knight-Ridder Newspapers is actively promoting its two-way videotex service, Viewtron, in south Florida.

Developed by Viewdata Corporation of America, Viewtron uses a home television set and telephone line and a videotex terminal from AT&T to connect viewers with more than 750 different topics. These include:

- News and Sports
- Home Banking
- Personal Finances
- Home Brokerage
- Education
- Books and Reference Material
- Travel and Entertainment
- Food and Dining
- Home, Family and Health
- Games
- Home Shopping

Viewers interested in home shopping review the products and services available—compare prices and features—and are then often able to place the order directly on their TV set. Over 150 merchants in more than 50 categories participate including:

- 10 Auto Dealers
- 12 Banks
- 6 Camera Stores
- 7 Electronic Stores
- 5 Flower Shops
- 3 Furniture Stores
- 6 Insurance and Financial Services
- 4 Men's Clothing Stores

- 4 Real Estate Agencies
- 4 Theaters
- 6 Shopping Centers
- 16 Travel Agencies

After a successful pilot test among 150 homes in Coral Gables, Florida, in 1980 and 1981, Viewtron was launched officially in November 1983 in three south Florida counties. Beyond Florida, if that system draws sufficient consumer interest, Viewtron would be expanded to other cities.

Although it will not start operations until at least 1986, three communications and marketing giants—CBS, IBM, and Sears—are teaming up to develop a national videotex service (Trintex) that will use personal computers rather than dedicated videotex terminals for home reception. It will emphasize computer processing functions as well as information retrieval and interactive banking/shopping/games/messaging.

It is interesting to note that in these ventures as well as in several others on the drawing board, there is involvement on the part of newspapers, magazines, and broadcast companies as information providers and co-venturers. In the "New Media World," it is obvious that these "Old Media" want to be a part of the action. The big question, of course, is how will people respond both to paying for a home electronic information service and using it. And how creatively and effectively will advertisers use it.

From an advertising standpoint, marketers can use videotex at any one of three different levels.

Level One—To Run an Advertising Message. A variety of advertisements of differing sizes can be developed and run in product- or service-related frames of information. Table 6-14 gives some examples.

Level Two—To Seek Direct Response. Products or services that have the potential of being ordered via the terminal, through an attached telephone handset, or through couponing in conjunction with a local retail outlet can use this avenue. While today's leading direct marketers may be expected to become leaders in the use of videotex for direct response, other advertisers can find it of value simply to offer free samples, coupons, or product information.

Level Three—To Supply Program Data. In this case the advertiser would supply not just advertising, but a full complement of informational programming. For example, a pet food company could provide the basic data on the care of pets. A diversified food company could provide information on a variety of recipes, or an automobile manufacturer could provide information on safe driving and car maintenance. The Shell Answer Man, for instance, could easily become a regular videotex feature.

Earlier in this book, I noted how the new media would impact not just on television, but on all media, including newspapers. Because of this, newspapers are experimenting with ways of effectively using cable and data transmission services to their advantage.

One example of such a joint effort is an Associated Press/CompuServe classified advertising videotex experiment. In 1982 the Associated Press, in conjunction with CompuServe, Inc., and 11 major newspapers, embarked on a test of a national network of a two-way interactive home information system in which subscribers could retrieve employment, real estate, and automotive classified ads. CompuServe is a computing services company that operates a nationwide interactive home information or videotex service with over 100,000 home subscribers in 1985.

TABLE 6-14 Possible Vehicles for Different Products

This Product or Service	Would Run In
Automobile	Travel Tips
Beer	Sports Scoreboard
Camera	Community Calendar
Financial Institution	Money Management Tips
Food Product	Menu Suggestions
Insurance Company	Medical Tips
Pet Food	Lost and Found Pets
Sporting Goods	Golf or Tennis Tips

If a marketer is to successfully use videotex and teletext as a part of his media mix, he should begin by understanding a few key guidelines for developing effective advertising for this new medium. BL Associates, Inc., is a firm concentrating on providing services targeted at educating and assisting advertising agencies in the field of videotex. Its director is M. Christopher Lockhart, and his book *The Advertiser's Place in the Evolution of Videotex*, provides some excellent guidelines in the form of "14 Golden Rules of Videotex." Many of the rules apply equally well to teletext.

1 Remember that practice makes perfect and early involvement allows the advertiser to discover what works and what doesn't work with his videotex advertising.

2 If an advertiser is serious about getting involved in videotex, it is best to acquire the services of a specialist in videotex and videotex advertising to avoid getting stuck taking too many false turns along the way.

3 Videotex is a highly dynamic medium and system improvements will continually provide advertisers with more useful tools to promote their products and services.

4 An advertiser unfamiliar with videotex should continually focus on his product or service and not be mesmerized by the technology.

5 Always evaluate videotex in terms of your company's objectives, goals, strategies, and tactics.

6 Because videotex can focus on precisely defined target markets, the advertiser should develop videotex marketing and advertising strategies suited specifically for each target market.

7 Videotex will allow advertisers to provide any quantity of information they wish to allow the consumer to make complete product comparisons.

8 Since videotex graphics lack the look of an actual photograph, advertisers should develop page identities that help the videotex users identify the advertiser.

9 Since the videotex user will access pages only to gain information, or to make a purchase, or to be entertained, remember your job is to inform—persuade—and *sell sell sell*!

10 Help the customer find your videotex data bases by making reference to it in other advertising.

11 It is better to provide too much information than not enough but be sure to index it for fast, convenient, and easy search and retrieval.

12 Videotex advertising should take advantage of the medium's interactive potential. If the consumer is involved in

some capacity, then learning is enhanced and a purchase is more likely.

13 Advertisers should make their videotex advertising a part of a valuable service that the consumer will associate with them and want to access frequently. The Shell Answer Man could become an interactive service answering specific consumer questions.

14 Since advertisements can be put on the system and removed at will by the advertiser, it is a good environment in which to test market new ideas and new product concepts.

TABLE 6-15 The top 25 direct marketers
Sales volume generated through direct marketing

Company	Sales volume ($millions) 1983	1982
Sears Merchandise Group	$2,092.0	$1,886.6
J.C. Penney	1,652.0	1,537.0
Montgomery Ward (Mobil Corp.)*	1,190.0	1,145.0
Colonial Penn Group	615.0	548.0
Spiegel Inc.	512.0	395.0
Fingerhut & Figi's (American Can Co.)	512.0	400.0
Time Inc. (magazines only)	385.0	320.0
Franklin Mint (Warner Communications)	378.0	410.0
New Process Co.	268.0	235.0
McGraw Hill Inc.	254.0	215.0
Columbia House (CBS)	220.0	200.0
Hanover House Industries (Horn & Hardart Co.)	215.0	140.0
L.L. Bean Inc.	205.0	192.0
American Express (TRS Co./merchandise sales only)	180.0	150.0
Herrschners, Brookstone, Jos. A Bank Clothiers (Quaker Oats Co.)	152.0	120.0
Avon Direct Products	144.0	121.0
Grolier Inc.	135.0	147.0
AT&T Communications	120.0	100.0
Jackson & Perkins, Harry & David's, Bear Creek (Bear Creek Corp.)	111.0	110.4
Wausau Insurance Co.	110.0	—
RCA Direct Marketing	110.0	100.0
World Book Encyclopedia (Scott & Fetzer Co.)	105.0	100.0
New England Business Service	103.0	91.6
Spencer Gifts (MCA Inc.)	101.5	85.0
Amba Marketing Systems (The MacDonalds Co.)	100.0	70.0

Source: DMA research department. Based on figures provided by participating companies surveyed Dec. 1983 to Feb. 1984. *Includes Signature Financial/Marketing Services direct marketing sales.

From Information to Shopping

Two basic formats to the video presentation of products and services have begun to emerge. Both approaches are represented by programs carried by MSN: The Modern Satellite Network.

First was the "Home Shopping Show," the longest running information program. The show offered 8½- and 30-minute segments of expanded information, demonstrations, and discussions of products and services. One program aired in November 1983, during which three companies introduced products to the cable audience:

- In a segment exploring ways to use Chex cereals to make nutritious and tasty snacks, Ralston-Purina introduced new Chex Oatmeal and Raisins.
- Briar Industries discussed Auto Wash, a time-saving device used to wash and clean cars, boats, motorcycles, mobile homes, and even houses.
- And Barbara Thornton, home economist with Universal Foods, introduced Red Star Quick Rise High Activity Yeast for faster baking.

To maximize the service orientation of the program, Ralston-Purina and Universal Foods offered viewers the opportunity to order product premiums containing recipes using company products. And Briar Industries provided a toll-free number through which viewers could purchase Auto Wash. A second MSN program, the Sharper Image Living Catalog, allows the viewer to purchase directly from home via a live video catalog.

A number of other information and shopping ventures are either in operation or in the testing stage. In one instance, the Entertainment and Sports Programming Network teamed up with Comp-U-Card, an electronic shopping service. "ESPN Pro Shop" costs $29.95 a year and offers savings of up to 40% on male-oriented sports and consumer electronic items. Members will receive a kit explaining how to use the service and describing the 60,000 brand name products from which to choose. When an order is placed, the Pro Shop provides the list price on the item. Members can either order it through Pro Shop or use the discounted price to bargain with local retailers.

In the future, it would not be surprising to see more cable networks expand the benefits they offer subscribers through similar efforts. A pay cable service, for example, could very logically offer a shop-at-home entertainment service, featuring audio and video equipment, games, home computers, and other related items.

Some very interesting applications of videotex in shopping malls are being tried out today. One is VideoTimes, operating in the Northpark Mall in Davenport, Iowa. More than 60 of the mall's 160 retailers use it to provide customers with product information and directions to their stores. Advertisements and information space is sold for the touch screen videotex directories located within the mall.

Your Imagination: The Only Limitation

More and more advertisers are experimenting with ways formerly reserved exclusively for print. Over the years, people have grown used to getting their news and information from the television set, and businesses are trying to capitalize on this.

Real estate firms are experimenting with the use of cassettes to "list" and "show" homes. The objective is to let prospective buyers screen out those homes they have no real interest in and save driving time for those houses that are real possibilities.

The first video cassette news release was taped in 1981 at a 45-minute press conference in New York announcing Pratt & Whitney's sale of a $600 million commercial aircraft engine. Within three hours of the news conference, the tape was edited and released by satellite to cable systems across the country.

Thus far, the electronic delivery of information via existing videotex services has lacked both high quality pictures *and* live action with sound. Quickscan Electronic Publishing of Burbank, California, has developed a system to overcome this. Textual and pictorial information is transmitted electronically over any standard television transmission system. It is received and recorded on a standard video cassette recorder located in the viewer's home, where it may be accessed and viewed on a television set at any time. Full-color pictures are delivered along with text, and stills can be intermixed with live-action and sound. The "pages" may also be "browsed" or scanned—by the viewer.

From an advertising standpoint, Quickscan is developing an electronic home information service "The Video Cassette Hour."

Consisting of helpful program information and long-form, information-oriented commercials, it will be broadcast nightly by satellite, transmitted to the home by local market television stations after normal broadcast hours, recorded on viewer's unattended VCR's, and played back at any convenient time. "The Video Cassette Hour" represents a marriage of several new media forms—satellite, VCR, and data transmission. It emphasizes how the new media may well evolve as amalgams or marriages of *existing* delivery systems.

In 1981, International Paper Company came out with the first video annual report. Its financial data and a video tour of its operations were highlighted in a half-hour cassette made available to stockholders in all tape sizes and aired on cable systems across the country.

A video trade show went one step further. The first such one was the "National Consumer Electronics Showcase" developed by Narrowcast Marketing. Such products as video disc systems, cassette

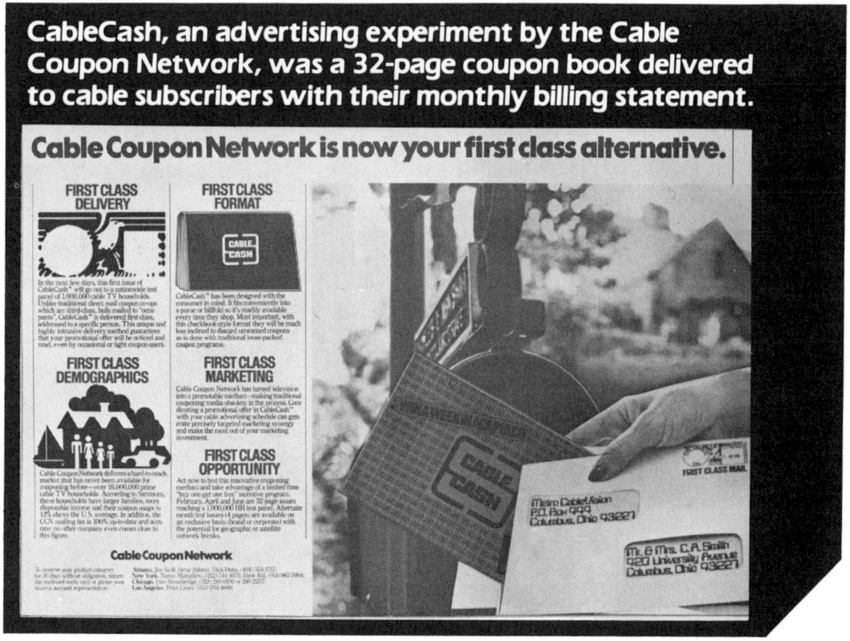

CableCash, an advertising experiment by the Cable Coupon Network, was a 32-page coupon book delivered to cable subscribers with their monthly billing statement.

recorders, electronic games, and home computers were demonstrated in 5- to 11-minute infomercials transmitted by satellite to approximately five million homes across the country. The messages were incorporated in an editorial environment that included updates on general product developments in personal communications systems, electronic games, advanced TV, home security, and similar items. The showcase provided a national video supplement to the regular live trade shows where advertisers regularly display and demonstrate their products.

The point here is that opportunities for the use of video to transmit informational advertising on all subjects, products, and services are basically limitless.

As a matter of fact, new media advertising can extend all the way to the bills that cable subscribers regularly receive for service. One thing that all cable subscribers have in common is the monthly bill they receive from their cable system. Coupons delivered to subscribers with their monthly statements can capitalize on this vehicle and can be an interesting addition to the marketing plan.

In a nutshell, today's advertiser is really limited only by his or her new media imagination. The Cabletelevision Advertising Bureau recognizes this, and to promote the benefits of cable to advertisers and advertising agencies, it created its first advertising campaign in fall 1983. The campaign played upon the theme of Albert Einstein's theory of relativity $E=mc^2$, which the CAB translated into "Effectiveness Equals More Cable!"

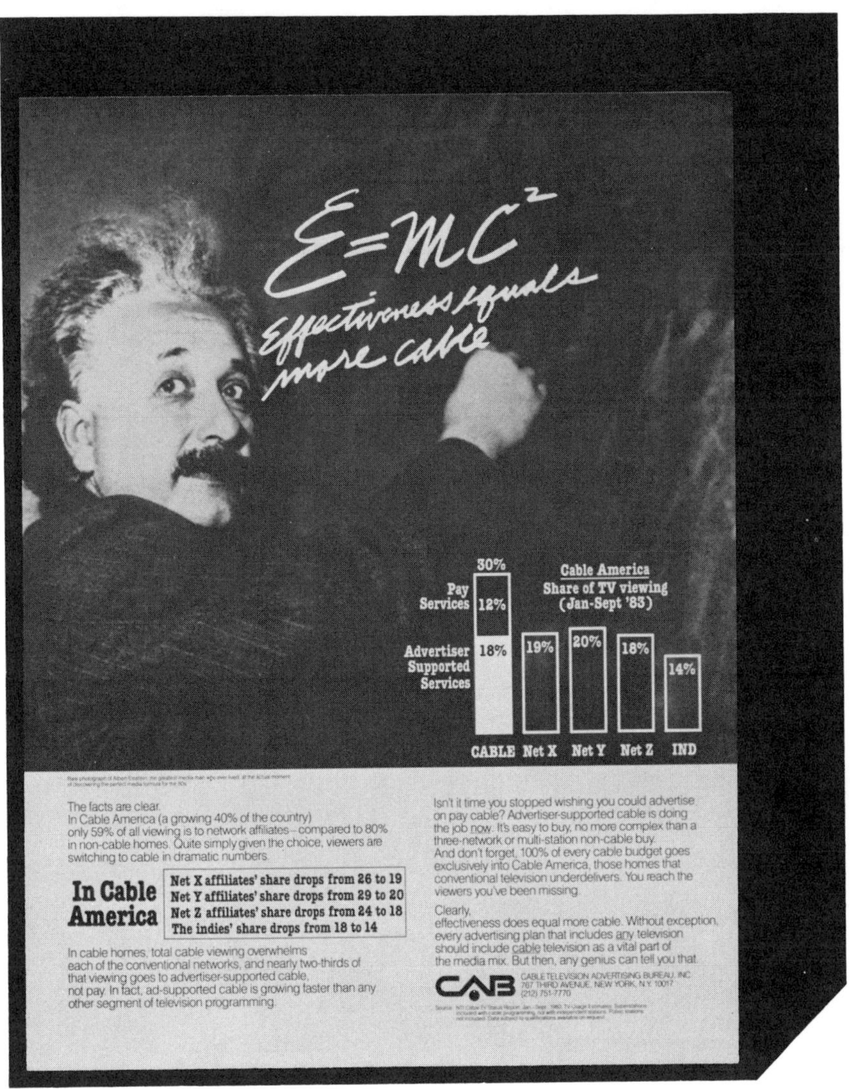
E=MC²
Effectiveness equals more cable
30%
Pay Services
12%
Advertiser Supported Services
18%
Cable America
Share of TV viewing
(Jan-Sept '83)
19%
20%
18%
14%
CABLE
Net X
Net Y
Net Z
IND
The facts are clear.
In Cable America (a growing 40% of the country)
only 59% of all viewing is to network affiliates—compared to 80%
in non-cable homes. Quite simply given the choice, viewers are
switching to cable in dramatic numbers.
In Cable America
Net X affiliates' share drops from 26 to 19
Net Y affiliates' share drops from 29 to 20
Net Z affiliates' share drops from 24 to 18
The indies' share drops from 18 to 14
In cable homes, total cable viewing overwhelms
each of the conventional networks, and nearly two-thirds of
that viewing goes to advertiser-supported cable,
not pay. In fact, ad-supported cable is growing faster than any
other segment of television programming.
Isn't it time you stopped wishing you could advertise
on pay cable? Advertiser-supported cable is doing
the job now. It's easy to buy, no more complex than a
three-network or multi-station non-cable buy.
And don't forget, 100% of every cable budget goes
exclusively into Cable America, those homes that
conventional television underdelivers. You reach the
viewers you've been missing.
Clearly,
effectiveness does equal more cable. Without exception,
every advertising plan that includes any television
should include cable television as a vital part of
the media mix. But then, any genius can tell you that.
CAB
CABLE TELEVISION ADVERTISING BUREAU, INC.
767 THIRD AVENUE, NEW YORK, N.Y. 10017
(212) 751-7770

CHAPTER 7

Creating and Producing Advertising for the New Media

When Marshall McLuhan wrote that "the medium is the message," he was obviously thinking only of television and not of the vast potential for advertising in the new media. Had McLuhan done so, he might more properly have said, "the medium is the message, *except* in cable, where the message can be the medium!"

Throughout this book, I have identified 10 major values of cable that must be considered in the advertising planning process:

1. Highly specialized programming allowing advertisers to zero-in on highly selective target audiences.
2. Low unit costs per announcement.
3. Ability to build a high frequency of exposure and compensate for loss of network television weight in cable homes.
4. Flexibility of advertising message lengths and forms.
5. Sponsorship opportunity with program identity.
6. Product exclusivity in programs.
7. Ability to test at low media costs.
8. Opportunities to tag advertising messages with direct response offers.

9 Ability to localize advertising for franchisees, dealer organizations, and wholesale and retail sales forces.

10 Creative commercial opportunities that match viewer lifestyles with program environment.

Jack Ziegler ©1983

Concentrating on Message Content

Until recently, most advertiser involvement in cable has concentrated on the medium and various approaches for using it to deliver existing broadcast commercials. It is equally important to focus on the advertising message that is delivered. In other words, cable's values must be brought alive through advertising creativity as well as through creative selection of the cable media vehicles.

The cable viewer tends to be upscale in education, occupation, and income. This viewer is a seeker rather than a passive receiver of information and entertainment and is attracted to the program content of cable in many of the same ways that a reader is attracted to the targeted editorial content of a magazine. As a result, there probably is less need for the advertiser to concentrate on attention-getting commercial devices than there is in broadcast television. The consumer will seek targeted cable channels of communication, so the advertiser is able to concentrate more on message content than on viewer contact. (See Figure 7-1.)

For example, a typical 30-second automotive commercial relies on very strong attention-getting devices to attract the viewer. There is a relatively short amount of time in which to deliver a high-information-content message, a two-minute "racer testimonial" developed by J. Walter Thompson USA stressed the selection of a Ford Mustang as the best car to race in the International Motor Sports Association American sedan class. "Inside Ford Racing" emphasized the merits of the suspension system and application of small-engine technology that pointed to a new breed of fuel-efficient race cars. Rob McFarlin, champion driver in the IMSA import sedan class, outlined how today's advanced Ford technology helped build better automobiles for the American consumer. The commercial, produced on tape at Road Atlanta, was designed to be placed in auto racing on cable. During the events, Rob was racing, giving the "testimonial" even greater impact. Since viewers have a natural interest in racing, the attention of the advertising message can be focused more on message content than on attracting viewer contact. In this case, there was a creative commercial opportunity to match up viewer lifestyles with program environment.

FIGURE 7-1 TV and Cable Communication

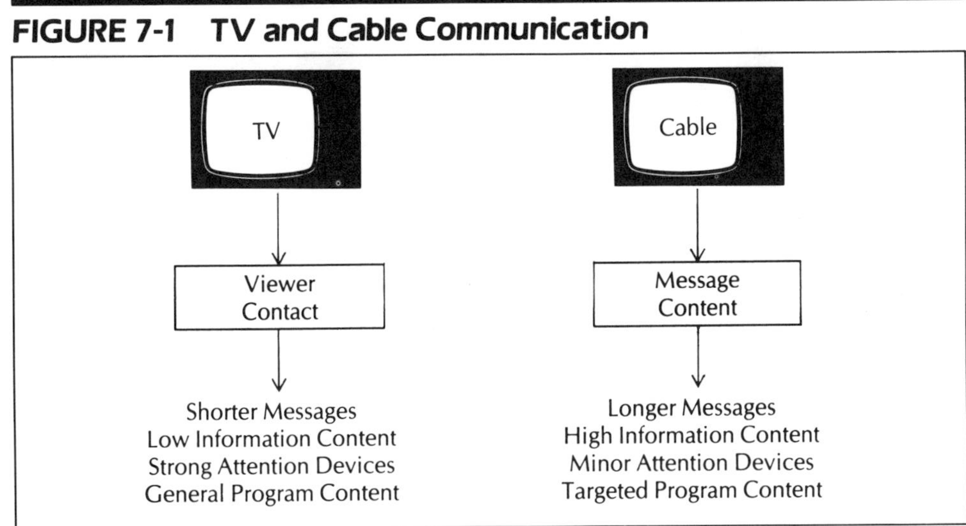

The Cable Creative Paradox: More Messages for Fewer Dollars

All this does not mean, however, that cable advertising can be "dull." Creativity is still vital, since viewers will always react far better to a well-executed 30-second commercial than to a poorly executed two-minute message. What will change is the approach of the message and the devices employed to communicate. The emphasis will be more on maintaining interest throughout the length of the message than on attracting initial attention to it. As a result, creative directors and producers will need to develop visual techniques that allow for the attractive and effective presentation of large quantities of information.

And, of course, they will have to do this at budgets substantially lower than for today's broadcast commercials. A $50,000, $100,000, or $500,000 commercial makes sense if the media budgets behind them are $1,000,000, $5,000,000 or $10,000,000. In cable, however, an advertiser will generally be running a large number of messages in one or more very low-rated programs with very low media costs. These cable messages will undoubtedly have a much shorter effective life than a broadcast commercial does since the reach of any individual message will be low and the corresponding wearout factor will be relatively high.

Wearout will especially be a problem for infomercials—in-depth cable presentations of information. In broadcast television, me-

FIGURE 7-2 The Cable Paradox

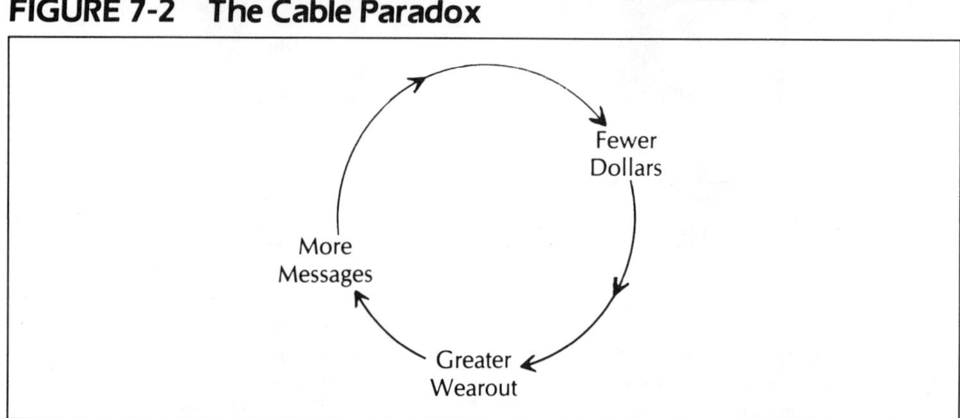

dia planners talk about the "effective reach" of a message and the need for a minimum of perhaps three exposures before it has been communicated sufficiently to produce a viewer impact. The upscale viewer, however, who is interested in a product or service and who seeks a specific channel, may actually be adequately reached with only one exposure of an infomercial. Similarly, the overexposure of an in-depth, long-form cable message may have a far more negative effect on the prospect than the overexposure of a 30-second broadcast message. (See Figure 7-2.) In other words, marketers will need to produce a greater number of cable advertising elements for substantially fewer dollars than they are accustomed to producing for broadcast television.

The Need for New Production Techniques

To meet the challenge of the Cable Creative Paradox, advertisers and agencies will be called upon to develop entirely new production techniques for the medium.

The average broadcast 30-second commercial cost over $70,000 to produce in 1984. In cable, however, creative directors and commercial producers must learn to develop messages of several minutes in length and bring them in at budgets of $5,000, $3,000, and, in some instances, under $1,000.

For the advertising agency, cable will undoubtedly result in new financial strains and needs for new compensation plans. Traditionally, agencies have been paid on the basis of a 15 percent commission on media purchased. Many agencies have fought the concept of flat fee compensation by clients for fear that fixed fees might not adequately cover variable costs.

Now, however, cable is totally changing the complexion of compensation. The small dollar amounts involved in early cable placements make it impossible for agencies to profit if they must develop special commercials via the commission route. Agencies will find that fees become a necessity if they are to be able to afford the time to produce effective commercials and campaigns for cable. As Paul Kagan, a specialist in the field of broadcast and cable consulting, pointed out: "Cable may prove to be an unexpected boon to ad agency profit and loss statements, not because of the springing loose of cable ad budgets, but due to the fall-out of fees after the cable detonation."

The New Media Message That Motivates

It must be recognized that the goals of cable communications and broadcast commercials are the same: the sale of a product or service to a prospective customer. One secret in the development of effective cable advertising approaches is to focus the maximum amount of creative attention on the *message content.* The three important terms in creating an effective cable commercial are the *media message that motivates* (i.e., sells), the *discernible product attributes,* and *significant consumer desires.* The discernible product attributes are those facets of a product's construction, appearance, and uses that will satisfy significant consumer desires—all of those reasons why people will want the product and buy it. (See Figure 7-3.)

In the development of a 30-second commercial, time limitations dictate that all attention must usually be focused on a single discernible product attribute that will satisfy a single significant consumer desire. Obviously there is not much time to accomplish this, especially when a portion of the message—often a substantial portion—must be directed at simply gaining and holding the viewer's attention. And of course, with interest being focused today on 15-second commercials, there is even less time to communicate with the customer.

Cable, however, offers the luxury of time—time to develop a *message that motivates*—during which an advertiser can show and discuss *many* attributes of the product (or service) that may satisfy *several* consumer desires. Just as in the case of the broadcast commer-

FIGURE 7-3 Producing the New Media Message That Motivates

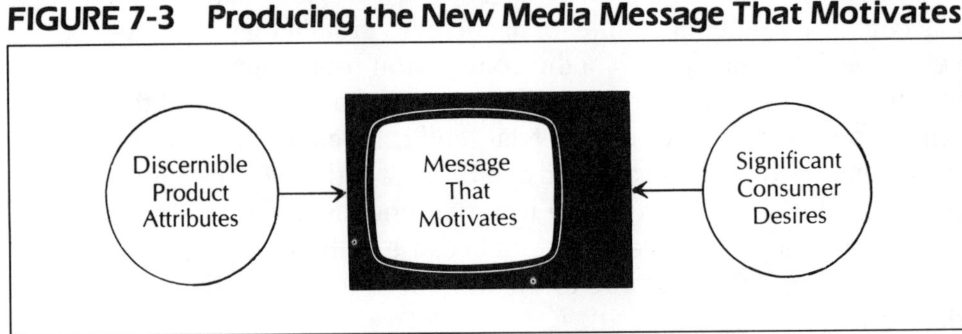

cial, the message must be interesting and memorable, but it can be interesting and memorable in cable by using simpler, more cost-efficient techniques.

Money-Saving Cable Creative Shortcuts

A major reason for the costs involved in traditional broadcast commercial production is the time needed (or taken) to develop an idea and execute the finished product. For example, a project might require 16 weeks from storyboard development to editing and post-production. (See Figure 7-4.)

For cable, there is the luxury of neither time nor money. Money-saving shortcuts begin with the selection of a production house that has probably had experience with retail commercials and/or industrial films. They will understand what it means when you want to deliver a package of five cable messages for under $25,000.

FIGURE 7-4 Traditional Broadcast Production Schedule

8 weeks — storyboard development by agency
2 weeks — legal approvals by agency, advertiser, networks

Selection of Producer

2 weeks — casting and location hunt
1 week — filming
3 weeks — editing and post production

16 weeks — from start to finish

In producing advertising for cable, you should:

1 Use relatively simple sets.

2 Use on-location shots, if convenient, but only if a minimum of make-ready is required.

3 Shoot several advertising executions back-to-back.

4 Use smaller casts.

5 Hold editing to a minimum.

And, on the subject of editing, it may be possible to use advertising material you already have in-house by:

1. Adding a tag to existing commercials.
2. Re-tracking the sound track of an old commercial.
3. Re-tracking *and* re-editing two or more commercials into one.
4. Cutting down and restructuring an existing industrial film.

The Quaker Oats Company did this when they sponsored the entire opening night of CBS Cable on October 12, 1981. Their advertising ran thematically through the evening much like a magazine.

In a segment on Mike Nichols's investments in Arabian horses, there was a Ken-L-Ration 83-second spot about a little lost dog. It utilized commercial footage shot for a television special 10 years earlier. Since the product label had been redesigned from when the commercial was first shot, a new tag was simply edited in at the end of the message.

A segment of Liz Swados's dance company, of interest to parents of young children, carried a Fisher-Price billboard plus a 104-second commercial taken from a half-hour Fisher-Price film showing how their toys were designed.

An Alec Guinness drama segment included a 38-second spot on the origin of the Quaker-man logo, again re-edited from a special commercial produced 10 years earlier.

Another way to hold production costs down is to produce multiple versions of a commercial, some for broadcast and others for cable use. For example, Chevrolet developed a special 90-second avant-garde hard rock music spot with visual collages to run on MTV: Music Television as well on USA's "Night Flight." But they also ran it on ScreenVision in movie theaters and developed 30-second and 60-second versions for late-night broadcast television. In a single production effort, three versions of an advertising message were developed for cable, theaters, and broadcast television.

Another example of editing and producing for cable involved "Kawasaki Sets You Free," a cable commercial produced by J. Walter Thompson USA as part of a developmental program to show clients the kinds of commercials that would be compatible with MTV: Music Television. The production included use of existing motorcycle footage, stock outdoor footage, and the development and recording of an original "heavy metal" rock and roll song. This two-minute message was produced as an experimental prototype to take advantage of "Life-Style Association" benefits of MTV. By using existing footage, production became largely an editing job with the completed project costing only $5,000.

Because of its low cost, cable will attract many "nontelevision" users who have no commercials. One example involved PantHer, a manufacturer of women's clothing. Dupont is a major fiber supplier to PantHer, and as part of its co-op advertising activities with them, Dupont and The Weather Channel developed a concept called Dress for the Weather. Clothing made by PantHer from fabrics with Dupont fibers provided the wearer with comfort under a variety of weather conditions, and programming related to today's weather was incorporated into the "Dress for the Weather" feature. Since PantHer was not a television advertiser, however, they had no commercials. The Weather Channel wrote and produced three commercials using one of their on-camera meteorologists relating product benefits (style and comfort) to her normal work environment (the hot lights of the studio and the cool lights of the computer room).

As advertisers investigate new and different ways to produce cable advertising that sells, they should seek services providing stock sound and video effects that can be integrated into their messages. Among the companies offering such services is Thomas J. Valentino, with a production music library of over 4,000 selections, a sound effects library of over 1,000 effects, and "Video Stock Shots" that include everything from a desert sunset to a forest in winter and from walking in New York to walking in space. Used skillfully, these effects can add to the impact of a cable message. If they are just "thrown-in" for effect, however, they may very well create a negative effect. The key to their use is skill.

Planning Local While Thinking National: The "Split-60"

In another chapter I will discuss cable advertising at the local level and how it may be bought, sold, and created. At this point, however, it must be emphasized that in planning and developing cable advertising at the national level, marketers should consider how these messages might be used locally for tie-ins with dealers, retailers, and franchisees. A "Split-60" commercial is one means by which the national advertiser can accomplish this. It represents two 30-second commercials that run back to back and create the impression of being a single message from a single advertiser. When run on a national satellite network, they appear as one unit, perhaps with two themes or appeals, from the advertiser. Local cable systems are advised of the schedule so that the advertiser's dealers, retailers, or franchisees are able to tie-in a specific localized message. Where this has occurred, the local advertiser's message runs over the second half of the national 30/30. In these cases, the national advertiser's first 30-second commercial stands alone without any creative impediment and leads into the local message.

The possibilities for these kinds of commercials are nearly limitless. For example, Century 21 might run a "Split-60" on ESPN. In markets where a local Century 21 office wanted to tie-in, it could override the last 30 seconds of the modular unit with its announcement. Or General Foods could feature some of its products in a "Split-60" on the Lifetime Network, and make available the last 30 for overriding by a local grocery chain. A *warning!* This approach will

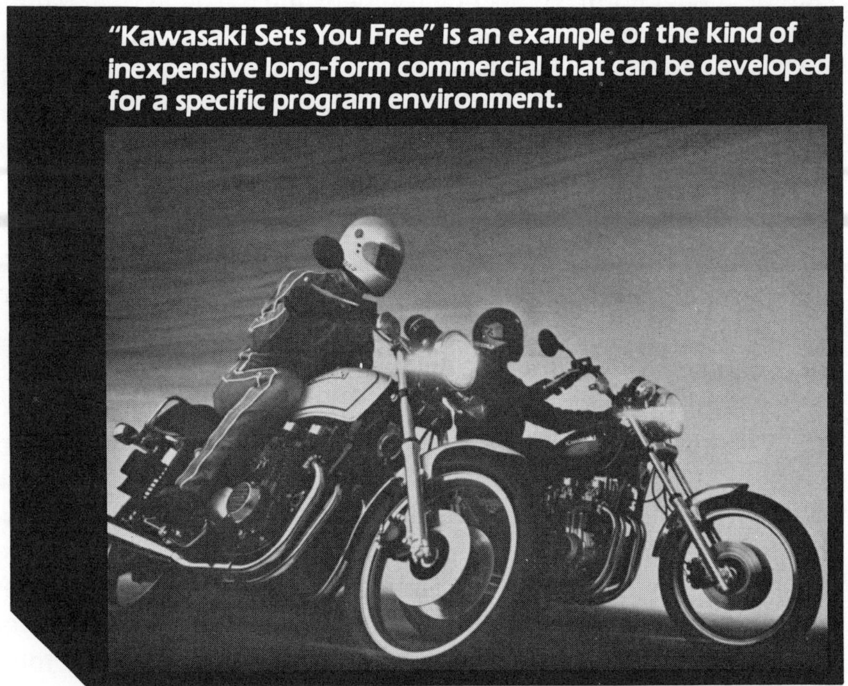
"Kawasaki Sets You Free" is an example of the kind of inexpensive long-form commercial that can be developed for a specific program environment.

become more common in the future as an increasing number of cable systems purchase the equipment to allow them to insert local commercials. Until then, its application will be limited.

Tips on 800 Numbers in Cable Communications

In developing advertising for cable, many companies have an interest in building a direct response mechanism into their messages. The Direct Mail Marketing Association can provide considerable insight as to how cable and 800 numbers can work together most effectively. Here are five of their suggestions:

1 Go to a good direct response advertising agency. It will know how to buy the time properly and to create a commercial that *sells.*

2 Properly utilize the 800 number in your ad. It should be displayed for at least 15 to 20 seconds.

3 Make sure there are adequate facilities to handle the 800 number, whether you are using an in-house facility or an outside telephone marketing service bureau.

4 Make sure that your 800 number has "rhythm" so that the number is easy to remember. For example, 800-228-8000 is far better than 800-647-2594 (and 800-CABLE85 is even better!)

5 Use a variety of 800 numbers for source coding. This will enable you to trace where your calls are coming from and to determine which cable effort is working best for you. This is especially important today with the very limited amount of available audience research data.

Understanding and Creating the Infomercial

Even though the infomercial is a relatively new cable concept, its roots reach all the way back to 1915 and the Pan-Pacific Exposition in San Francisco. In its exhibit, General Electric Co. presented an 18-minute silent film displaying the first all electric home, complete with electric vacuum cleaners, heaters, toasters, irons, washing machines. . .and even an electric cigar lighter shaped like a telephone. This film has been edited down to nine minutes and is on display at the Chicago Historical Society. It is a classic infomercial created in the days when cable was a message you sent from New York to London.

What is the infomercial? A broadcast commercial seeks to generate awareness for a product or service, to build brand identity, and to register usually a single major sales point. In contrast, an infomercial, through the luxury of time, can explain the product's or service's benefits, uses, and characteristics. It can also provide interested viewers with the means of securing more information, a sample, or the product or service itself. As Bill Harvey, publisher of the *Media Science Newsletter,* pointed out, "By involving the viewer more in the message, an infomercial can be regarded by the viewers as more relevant, absorbing, sincere, and real than a short-form broadcast commercial."

All products and services were not meant to be sold in the confines of 30- or 60-second television messages. Unfortunately, the structure and pricing of broadcast television has generally necessitated this. Infomercials, however, allow the marketer to break out of this mold and provide the more detailed selling information he would like to give in broadcast television but cannot afford. And, because of the selective audience appeals of many cable networks or local system channels, he can also deliver the message to a more targeted rather than to a mass television audience. In a nutshell, infomercials can provide advertisers with the in-depth communications ability of print, coupled with the video impact of television. (See Figure 7-5.)

FIGURE 7-5 The Infomercial

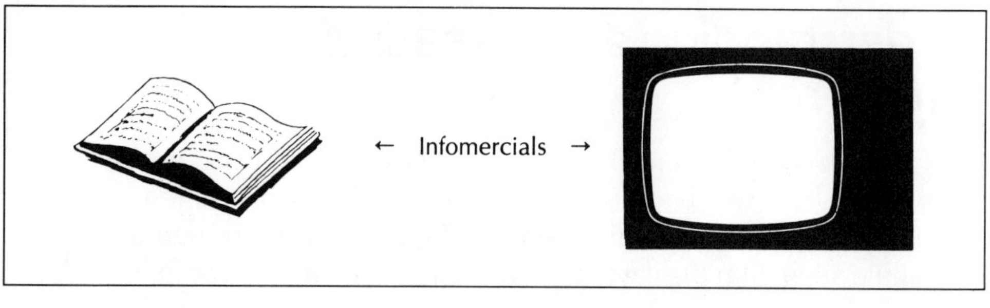

← Infomercials →

Producing Effective Infomercials Efficiently

While a 30-second broadcast commercial may cost well over $100,000 to produce, an infomercial can be delivered for well under $5,000. Its aim is to impart information, and this helpful service aspect, rather than expensive production techniques, should hold the viewer's attention. Costs are held down by shooting several infomercials on a single day (rather than one over several days), using stock footage where available and applicable, editing on the spot, using tape and not film, and employing a minimum number of cameras *and* a minimum number of creative and production people.

At the local level, an infomercial would be even simpler to produce if modular material supplied by the national advertiser was used and/or if commercials were shot on location at the local retailer, dealer, or franchisee's place of business. In this case, it might, for example, cost only $100 to go into the local lawn and garden store and shoot an infomercial on how to select plants for your yard.

Unfortunately, an infomercial can have fast viewer wearout and limited audience reach. Therefore, an advertiser might find it necessary to produce more of them than he would 30-second commercials for a broadcast pool. This is another reason why the cost of producing each infomercial must be held down.

Today, most advertisers placing 30- and 60-second commercials on cable satellite networks are simply running existing broadcast commercials, something that is not particularly innovative. However, if advertisers can do helpful three- to seven-minute cable infomercials, they may perhaps be able to do effective, interesting *30- and 60-second* mini-infomercials for broader cable use. (Some of these could even be utilized on broadcast television—perhaps in the day-time and fringe areas. Considering the volume of advertising that advertisers place in these dayparts—and the total flood of commercials there—these new-form messages might be very well received by viewers.) With the trend in commercial production being that the more you spend, the better the message, *it would also be very worthwhile to begin testing new commercial forms using new, less-expensive informational production techniques.*

Guidelines for Creating Infomercials That Sell

Research has generally indicated that infomercials should be well received by viewers. They are very receptive to the idea of helpful, informative, long-length messages and often point out that they would like the kind of buying information that "old-time salesmen" used to provide. Unfortunately, salespeople today generally aren't what they used to be, and perhaps that is why there may be this appetite for video infomercials. One warning: don't think that all you have to do to produce an infomercial is stretch a 30-second commercial to two minutes or so. Viewers will recognize this immediately as nothing more than a long-form commercial and will be turned off.

Here are some guidelines for the creation of effective, selling infomercials:

1 Remember that the job of the infomercial is to provide depth of impression, not merely broad exposure of a message. It must provide the viewer with genuinely helpful information.

2 The infomercial gives you the time to tell the viewer what you are going to tell him about, to tell it to him, and then to tell him what you have told him!

3 Be sure to explain to the viewer how to go about getting more information on the product or service. *And* make it easy for him to do so.

4 Provide a built-in direct response mechanism where appropriate.

5 Avoid exaggeration and puffery.

6 For added effectiveness, consider using short-form, tie-in commercials that can run on cable networks or individual cable systems and reinforce specific points in your long-form informational messages. For example, five- to seven-minute infomercials by a tax service like H & R Block could provide in-depth information on how to maintain tax records and assemble your material each year at income tax time. A series of 30-second "mini-messages" could focus on separate tips on how to maximize your savings in specific areas. (See Figure 7-6.)

7 Unlike broadcast commercials, which may be exposed up to 10 times without any deleterious side-effects, infomercials tend to wear out fast. A single exposure may be sufficient to convey its intended message. Hence, if an ad-

FIGURE 7-6 Infomercials: Long and Short

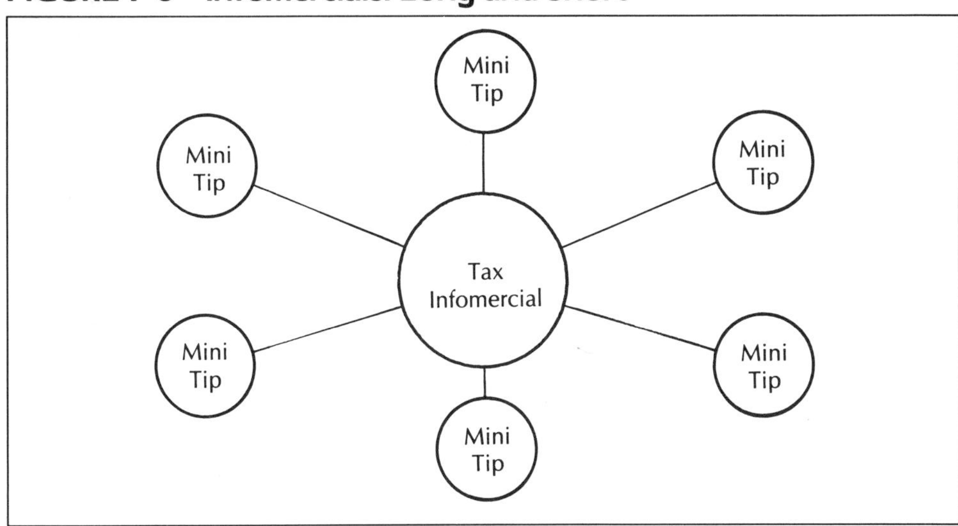

vertiser intends to use infomercials over an extended period of time, it is necessary to produce a sufficient number to avoid undue wearout of any single message. The problem is further exacerbated by the relatively low reach of an infomercial against its selected cable audience relative to the broader reach of a mass-audience broadcast schedule.

8 An infomercial should have a spokesperson. He need not be an actor and preferably should be someone close to the product at the client company. If the product manager, R&D manager, or even the president of the company can present well, he or she will make a very creditable spokesperson for the message.

9 If you have produced an infomercial, extend its life by making it available for showings at sales meetings and distribute it on tape to dealer organizations.

An excellent example of the development of an effective infomercial involved Sealy, the bedding manufacturer, and the Modern Satellite Network's "Home Shopping Show." A half-hour infomercial was produced in three segments that could be used together or shown individually. This consumer information program included Sealy's director of medical research, the director of the Center for Pain Studies at the Rehabilitation Institute of Chicago, the director

of product development, and the director of marketing. Sealy had "freedom of time" to discuss the importance of selecting quality mattresses to ensure proper rest, good health, and a "sense of wellness." Obviously, none of this could have been done in only 30 seconds.
An example of an informative print vehicle that could logically be translated into an informative cable vehicle was Campbell Soup Company's "Better Life Guide #1," which appeared as an 8-page insert in *Reader's Digest.* The booklet was the first in a series offering helpful information on health and fitness. It included discussions of:

- The importance of physical fitness
- Measuring how well your cardiorespiratory system and muscle system perform
- Sustained weight control and a balanced diet
- Regular exercise

Together these topics would be easily translated into a cable program on health and fitness. Or, individually, each of them could become a three- to seven-minute infomercial.

Creativity and Computers: An Informative Selling Approach

The growing interest in home computers is evidenced in the television series that premiered on PBS stations beginning in fall 1983. The "New Tech Times," hosted by former FCC Commissioner Nicholas Johnson, was a weekly news magazine. "Making the Most of the Micro" was a tutorial program imported from the BBC. And a 12-part series, "Bits and Bytes," was produced by New York's WNET and ran in 10 U.S. cities.

Cable's capacity to deliver in-depth, informative advertising messages for often complex products and services would seem to have considerable appeal for the computer industry. With the enormous changes taking place in the field, the need for consumer education is greater than ever. Potential customers are confused and frustrated. In their hearts they may want to invest in a computer system now. But in their heads they may be saying:

- Maybe I should wait.
- Prices may come down more.

- In a year I'll be able to buy a better system.

A computer company or computer store could become "The Computer Counselor," providing good and helpful information to customers and potential customers on the entire computer selecting-buying-using process.

- Kraft has done this in the food area through food preparation suggestions.
- Shell has done this with Shell's Answer Man.
- Henry Block has done this in the tax field with his "17 Reasons Why H&R Block. . . ."

The cornerstone of "The Computer Counselor" would be a series of half-hour programs covering:

- Computers and Home Management
- Computers and Kids
- Computers and the Small Businessman
- Desktop Computers
- How Much Capacity is Enough
- Getting Started. . .A Guide for the Novice
- Software for the Hardware
- Troubleshooting the Computer
- Designing Your "Computer Room"
- Computers and Financial Management
- Questions to Ask When Buying Your Computer
- The World of Word Processing
- Computers and Games

Since the program is in large part the commercial, actual commercial breaks could be limited! Placement of the series would be on one or more of the cable satellite networks (CBN, SPN, MSN, The Learning Channel).

Short (60 second) mini-information spots could provide added frequency of exposure. The series would be cross-promoted in print and viewers could request brochures with further information. Tapes would be shown on monitors in computer stores and would be provided to schools and adult education classes. The concept is a blend of cable video, public relations and direct response. And while national in scope, the related efforts would all be built up at the local level.

Tapes, Discs, and the Future

As we move from the mid to the late eighties, attention will be focused not only on cable, but on the development of appropriate forms of advertising on video tape or video disc. Service/product areas such as baby and child care, gardening, exercise and fitness, travel, home computers, and photography are just a few examples of areas in which there is a high degree of either general or specialized consumer interest and where advertiser-related tapes or discs might be meaningful commercial vehicles. Bill Harvey offers five excellent tips on producing tapes and discs that should be considered by anyone investigating this media approach:

1. It should not be something you can get on television.
2. It should require or encourage re-watching.
3. It should look good on wall screen as well as on conventional televisions.
4. It should appeal to movie and record buffs as well as videophiles.
5. It should be produceable without an enormous movie budget.

The point is that *all* of the new media can and probably will be advertising supported to at least some degree. It will require imagination and creativity to develop advertising that will most effectively communicate a marketer's messages, regardless of whether they are carried via cable, cassette, disc, videotex, or any distribution system not even yet thought of. This will be the advertising creative challenge of the years ahead.

CHAPTER 8

Measuring Results: Qualitative and Quantitative Approaches

The Need for *Facts*. . . Not Just *Faith*

The dramatic growth of the new electronic media has been accompanied by an increasing interest in cable advertising and by an increasing interest in hard, actionable research data. Until now, many advertisers have been willing to spend their dollars in cable largely on a matter of faith that they would receive in viewer response a value approximately equal to the money they have invested.

This is in sharp contrast to the buyers of magazines, newspapers, outdoor, radio, and television. They evaluate their media purchases on the basis of data from MRI, SMRB, ABC, Arbitron, Nielsen, TAB, and many other research and auditing firms. Today, however, as cable advertisers consider larger investments in the medium and as many of them investigate extending their involvement from a national to a local basis, they must justify these expenditures of their money. It is no longer just experimentation. They need confirmation of the viewing audiences.

The cable research needs of advertisers were highlighted in a survey of the nation's 300 top advertisers, researchers, and agency media buyers and executives in late 1983. They were asked about their key broadcast and cable research interests and concerns. (See Table 8-1.)

TABLE 8-1 ARF Survey of Broadcast and Cable Research Concerns

		% of Respondents Citing Interest or Concern
1.	Audience measurement with people metering	68%
2.	Data for planning media mix decisions	50
3.	Effect of clutter on audience response to ads	49
4.	Audience fragmentation	48
5.	Editing/zapping of commercials	48
6.	Viewer/audience response to specialized channels	44
7.	Effect of program environment on audience response to ad	43
8.	Integrity of ratings from national meter samples	42
9.	Integrity of ratings from local meter samples	42
10.	Channel-switching behavior	42

Source: Paul Kagan Associates (December 1983)

From cable's standpoint, the survey pointed up the importance of accurate measurement of who is watching what and the impact of programming and program environment on advertising effectiveness. A separate study of 250 cable advertising users, conducted by media consultant Jack Myers, showed advertisers felt cable needed much more data to support its sales efforts. Overall, 43 percent of those surveyed believed research other than household ratings should be used for buying cable advertising. And 81 percent called specifically for demographic data.

What Is Available Now

Until early 1983, a very large void existed in the availability of cable audience measurement data. While there is still no single measurement service covering the entire medium on either a national or local basis, what once was a very parched desert is beginning to sprout a number of needed oases.

- Nielsen meters the viewing of 1,700 sample households around the country. When a cable network reaches 15 percent national coverage (about 13 million homes), Nielsen can provide metered measurement on it.

- In terms of specific advertiser-supported satellite networks, WTBS was the first to be regularly measured by Nielsen's national meter service and was joined in the summer of 1982 by the Cable News Network. By early 1985, 10 additional networks had joined the group.
- Both Arbitron and Nielsen's Home Video Index can provide special studies of satellite network and local system audience delivery in specific markets using telephone coincidental surveys. Marquest Cable Ratings and Information & Analysis' Local Cableratings also conduct customized local market cable rating studies.
- Throughout the year, Nielsen provides monthly reports on trends in cable and pay cable usage overall.
- Audits of Great Britain (AGB), one of the largest marketing research firms in the world delivering metered rating data in England, Germany, Italy, Holland, and Hong Kong, is testing its "people meter" in the United States. Nielsen and Arbitron are also testing people meters to determine if they represent a practical, accurate alternative to current methods of collecting viewing data.
- Information Resources Inc., the market research firm whose BehaviorScan service uses scanning equipment to measure supermarket product movement, has developed a viewing tracking service using electronic monitoring equipment.
- Qualitative profile information on cable's general characteristics is available from MRI, SMRB, NPD Electronic Media Service, SRI Cable Study, and other omnibus services. (See Table 8-2.)
- A service to evaluate television and cable's program appeal and impact levels is being developed by Television Audience Assessment of Cambridge, Massachusetts.
- Another service aimed at measuring the qualitative appeal and familiarity of programs, movies, and specials among basic and pay cable audiences is "Cable Q Ratings," produced

TABLE 8-2 Who Views Cable?

Demographic Comparison of Adults in Cable and Noncable Households

	Total U.S.		Cable Households	Non-Cable Households
Age	%	Index	Index	Index
18-34	40.6	100	103	99
35-54	30.6	100	114	94
55+	28.9	100	81	108
Income				
$40,000+	17.8	100	124	90
$30,000-$39,999	17.9	100	120	93
$20,000-$29,999	24.8	100	107	97
$10,000-$19,999	23.0	100	89	105
Under $10,000	16.5	100	60	117
Education				
Attended/Graduated College	31.8	100	108	96
Graduated HS	38.6	100	105	98
Did not graduate HS	29.7	100	84	107
Employment				
Mgr/Prof	17.9	100	120	91
Other employment	42.3	100	107	97
Not employed	39.8	100	84	107
HH Size				
5+	18.7	100	108	96
3-4	39.9	100	114	94
2	29.8	100	90	104
1	11.6	100	65	116

by Marketing Evaluations in Port Washington, New York. The "Q Scores" are especially useful in assessing the potential strength of relatively new shows which have not had enough time to build a substantial audience.

In addition to cable measurement, advertisers and researchers obviously have an interest in data to evaluate the usage patterns of other electronic video alternatives including video games, video shopping, video tape and video disc systems, and video information services. However, because cable is today the most developed new media form with the greatest number of advertising opportunities, the major focus is on more fully building its research base.

Purchasing Power of Cable Households

	Adults in Cable Households Index	Adults in Non-Cable Households Index
Own American Express Card	117	93
Use Travelers Checks	110	96
Take Business Trips	118	92
Stay in Hotel/Motel	112	95
Rent Car	119	92
Purchase New Car	111	95
Acquire New Car Loan	133	85
Acquire Mutual Funds	120	90
Acquire IRA/Keogh	117	92
Own Video Recording Equipment	126	89
Own Video Games	128	88
Purchase 35mm Camera	119	92
Purchase Electric Clothes Dryer	125	88
Purchase Microwave Oven	127	87
Purchase Electric Power Mower	125	87
Spend over $50/Year in Mail Order	114	93
Drink Low Calorie Beer	117	92
Drink Domestic Wine	114	95
Are Heavy Users of Fast Food Restaurants	112	95
	Female Homemakers in Cable Households Index	**Female Homemakers in Non-Cable Households Index**
Spend Over $100/week on Groceries	121	91
and Are Heavy Users of:		
Cold Breakfast Cereal	112	95
Paper Towels	111	96
Moist Pet Food	119	92
Fabric Softeners	120	92
Disposable Diapers	124	91
Plastic Garbage Bags	123	91
Potato Chips	117	93

Source: MRI, Spring 1983

What Is Needed

The significant point is that audience research data are available today on over 150 consumer magazines, newspapers in some 50 markets, and all broadcast radio and television stations.

What national advertisers and agencies are seeking is comparable data on all of the advertising-supported satellite cable networks utilizing the same data bases and coming out several times a year. And at the local level, cable operators and marketers interested in the placement of local cable advertising need cost efficient methods of securing some form of audience research on more than just a sporadic basis. The big question is, How does the industry get from where it is to where it should be?

A Question of Technique: What to Measure

Back when the earth was flat and people didn't know what lay beyond the horizon, everyone was afraid to take chances and venture into the unknown. Then one day we learned the earth was really round. Everyone wanted a piece of the action, and Hertz Rent-A-Boat could barely keep up with the demand. Explorers went out looking for something, and many were not even sure what they were seeking.

It's now 500 years later, and we're still in the era of exploration. This time, however, instead of looking out over an ocean, we're treading water in a sea of cable confusion. On one hand, we're trying to come up with ways of measuring the audience of cable while, on the other hand, only a small number of explorers really understand what the medium is all about.

The initial problem is really not how to measure cable. It's *what* to measure. The problem is to get our explorers' crews to understand that cable is not just a lot of additional low-rated television channels, but in many respects an entirely new medium. (In other words, the cable world is not flat. It's round. And, in fact, as scientists discovered about the earth, it may even be oval!)

Cable gives us an opportunity to decide exactly what is out there we want to measure before we start counting heads.

Learning from the Past

Too often in the past, research has dictated how we should evaluate media rather than the reverse. Just ask any media planner:

1 Why do we talk about four-week television reach?

2 Why do we examine radio on a one-week cume basis?

3 Why do we compare magazines on a total audience basis?

Many either won't know or will think that it's based on knowledge of how marketing and media influence each other. In reality, the answers are that:

1 Nielsen initially had six analysis periods each year. They picked the four center weeks of each two-month period to report on. That's how we got four-week reach.

2 We examine radio on a one-week cume basis because diaries are kept for a week.

3 Total issue audience became gospel largely through the efforts of *Life* magazine to come up with big numbers.

The problems in magazine measurement should be a lesson as we look ahead to measuring cable. We have modified through-the-book, recent reading, modified recent reading, calibration, and cover recognition, and only now are we really acknowledging that all of them may be right and that they just may be measuring different things.

In the past, we have forced media into existing measurement techniques, expecting the measurement devices to be something they are not. We must avoid being mesmerized by today's measurement hardware. We have a clean slate with cable, and the first task must be to understand just what it is and what it is capable of doing before we try to measure it.

Reflecting How Cable Is Used

From the outset it is apparent that any cable audience research methodology must reflect how people view cable and how advertisers should buy the medium. For example, cable networks or local systems might sell by dayparts or on the basis of individual programs or combinations of programs. In some instances, audience data that report daily or weekly cumulative circulation levels across dayparts might be sufficient. In other cases, where cable networks or large interconnects reached very large numbers of homes and where commercial time is purchased much as it is for broadcast television, individual rating information is needed. And, in still other instances, there might be more interest in general qualitative data on the characteristics of the audience, similar to that available for newspapers and magazines.

"DRESSES, FOOTWEAR AND LAWN EQUIPMENT CLOBBERED THE HULK, HARPER VALLEY PTA AND THE ABC MOVIE!"

A First Step—Counting Cable Households

The Problem of a True Count

A continuing problem in cable measurement is simply getting an accurate count of the number of cable households in existence. At any given point in time, in any given area, there may be anywhere from a 10 percent to 20 percent difference in the number of cable households, depending on whether the data come from subscriber counts or actual house-to-house censuses. The differences are due to:

- Piracy. Households are buying illegal convertors and hooking up to the cable themselves.

- Homes that have cancelled their subscriptions, but have not yet been disconnected.

- Large apartment and condo complexes where the building management does not report the full number of individual unit subscribers.

- Homes that have subscribed but have not yet been included in the subscriber count.

Some experts estimate that as many as 1 out of every 10 cable homes is currently receiving its signal illegally by hijacking signals with homemade antennas, by buying illegal convertors and bypassing in-home decoder boxes, or by climbing telephone poles to knock out security traps.

Obviously, a problem in cable audience measurement begins simply with knowing how many cable homes are out there.

From Cable Count to Coverage Area

Before one attempts to measure the viewing audience of specific cable networks or local cable systems, it is necessary to establish a more basic audience measurement that is just one step removed from the cable household count. This is the "coverage area" of the potential cable advertising vehicle. It represents those cable homes located in areas that can receive the cable service. Advertisers will want to examine the distribution of homes served by a cable network or cable system in relation to their own sales territories or marketing areas. For example, before the Cable News Network was able to supply advertisers with metered Nielsen audience data, it provided information on the distribution of their household count by Nielsen territories and county size. From these data, an advertiser could see where the service's audience was concentrated and if it was distributed in line with population or other measures.

The Weather Channel used data on the distribution of its subscriber households to demonstrate to advertisers that 84 percent of its viewers lived in the most populous A and B counties with the greatest buying power. (See Table 8-3.)

TABLE 8-3 The Weather Channel Viewer Distribution

	The Weather Channel	All Cable Households	All TV Households
A Counties	45% }84%	28% }64%	41% }71%
B Counties	39%	36%	30%
C Counties	12%	22%	15%
D Counties	4%	14%	14%

Source: The Weather Channel, Fall, 1983

Advertisers should be very careful when it comes to examining cable household distribution or audience data within customary television universes such as ADI's or DMA's. Arbitron's Area of Dominant Influence (ADI) and Nielsen's Designated Market Area (DMA) were developed in the 1960s to define areas with homogeneous television station availability. The entire country was divided into approximately 200 ADI's or DMA's, and each area reflected all of the counties that devoted a majority of their viewing time to stations in the home county. The New York ADI, for example, encompasses 30 counties.

Bill Harvey has developed a concept of the Area of Community Influence (ACI). It follows the lines of the cable franchise area so that a given ADI might contain as many as 100 different cable systems. The ACI concept not only allows an advertiser to determine how many cable systems must be bought to cover a given marketing area, it also more accurately pinpoints the homes that would be reached by buying cable on a specific number of individual cable systems.

Some Basic Problems Facing Cable Measurement

Over the years, a number of methods have been developed to measure television viewing audiences. The most common methodologies have involved diaries, telephone interviews, and meters. In broadcast television, the cost of these services generally has been shared among several stations in a given local area. Because cable operators have exclusive franchise areas, however, the expense of audience research for a single local system must be borne by that system itself and cannot be shared. In addition, cable, both on a local and on a national basis, represents a medium of many more channels than in the case of television and with far lower ratings. Furthermore, audiences of most of the national satellite networks are distributed in a non-uniform pattern across the country, often making it extremely costly to develop a sample of viewers. And, of course, at the local level, many system owners are simply totally unaccustomed to research, have not involved themselves in the sales of local advertising, and must be convinced that it is to their benefit to provide audience data.

Taken together, all of these factors point to the need for the development of methods of reliable, low-cost audience measurement that can be effectively and efficiently used by both advertisers and the cable industry in buying and evaluating cable advertising concepts.

Five Basic Audience Measurement Methods

Five methods presently exist for measuring cable audiences—diaries, meters, telephone coincidentals, telephone recall, and two-way interactive cable.

In the long run, two-way interactive cable will provide instantaneous feedback as to exactly what cable channels are tuned and in what homes. It may be many years, however, before a sufficient number of them are distributed throughout the marketplace to provide the needed sample frames. And, of course, even with a two-way interactive response device that does not rely upon reporting by an individual, there is the problem that it measures set tuning rather than individual viewing.

While meters may well emerge as the most effective means of measuring the audience of national satellite networks, they, like two-way interactive cable, are not 100 percent adequate. They do not currently reflect who is viewing the cable service, although tests are underway by Arbitron, Nielsen and Audits of Great Britain to develop a "people" as opposed to a "household" meter. And, from a local standpoint, individual meters are extremely costly. Today, they are used for local broadcast measurement only in a limited number of markets, and it is doubtful that they will ever expand to the measurement of individual market cable data.

Telephone recall techniques are used to gather data on viewing anywhere from today to yesterday or to the previous seven days. In general, the longer the recall period, the greater the influence of memory on the accuracy of the response. Hence, it is generally regarded that recall is of greatest value in collecting data on audience levels over a period of time rather than on individual quarter hours.

Since memory is not involved in telephone coincidental surveys, they generally have been regarded as the most accurate method of gathering data on who is viewing now. From a cost standpoint, however, very large samples are needed, since every time period measured requires a sample sufficient from which to project results.

From an economic standpoint, diaries are a very efficient method of measurement and there are a variety of approaches to it. There are household diaries and personal diaries kept by the individual members in a home. Some diaries measure an entire week's activity and others measure only a single day's worth of viewing. Unfortunately diaries rely on the diarist's memory, and the element of memory has typically been responsible for showing cable audiences viewing levels that fall well below those reported either by meters or by coincidental methods. In broadcasting, it has long been acknowledged that diaries tend to underreport the audiences of lower-rated fringe and daytime programs as well as audiences of independent TV stations, ethnic stations, and public broadcasting stations.

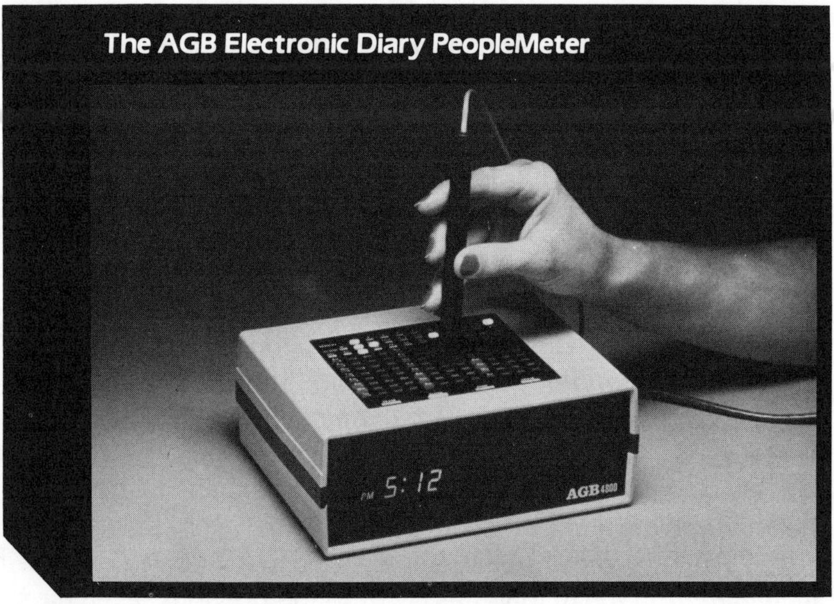

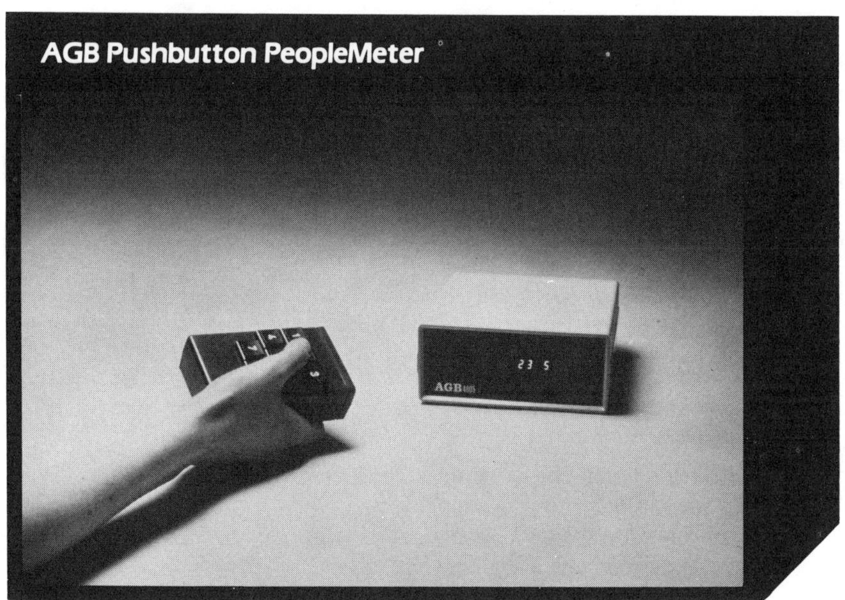

With these problems, some might ask, why has the diary been used? The answer is simply that it has been the most economical method to gather large quantities of viewing data for an entire week from an individual household.

Any decision as to which cable research method to use must take into consideration the cost of the method used to get the data generated. As might be expected, the more accurate the data generated, the more expensive the technique. (See Table 8-4.)

The Need for Cable's *Own* Measurement Systems

Because cable audiences are so fractionated and the numbers far smaller than for broadcast television, it is generally acknowledged that special studies are required to measure them. They cannot be adequately picked up in existing Arbitron or Nielsen television rating reports. While both Arbitron and Nielsen have developed ways of reporting broad market audience totals of cable audiences in their regular television reports, these generally are insufficient for evaluating the medium, and, in most cases, very little viewing data show up. This is because the cable audiences are related to an entire television coverage area rather than to only their own subscriber base.

TABLE 8-4 Cable Research Comparograph

Methodology	Description	Data Yielded	Approximate Cost (1984)
Interactive Cable	Requires diaries placed in field to determine demographics behind specific viewing.	Tracks exact TV set usage through census or sample.	Cost of scanning device, computer software and, if required, an impartial third- party audit.
Meters	Requires diaries placed in field to determine demographics behind specific viewing.	Tracks exact TV set usage of sample	In excess of $350,000 a year for 500 meters.
Telephone Recall (24-hour)	Asks additional questions for programming research. Relies on participant's ability to recall viewing.	Daily cume for each channel by daypart, and weekly cume for each channel. Demographics broken down into sex/age categories.	Starts at $3,000 for 200 completed phone interviews over a seven-day period.
Telephone Coincidentals	Simple formula of questioning household on viewing at exact moment of phone call. Accurate but limited information.	Average quarter-hour ratings and distribution of audience in terms of men, women, teens and children.	$400-$500 for over 200 completed phone calls to measure particular program or channel.
Diary	Follows household viewing across a week's time. Assertion is that method understates programming of limited viewing.	Average quarter-hour ratings, weekly cume data, and numerous demographic breakdowns.	$9,000-$10,000 for 250 diaries (price will differ if tabulations are made from existing diary base for broadcast research).

If, for example, a cable system covers 20 percent of a total market, audience data for the system must be measured and shown as a percentage of its own individual franchise area, not as a percentage of the entire television marketplace. (See Figure 8-1.)

The general consensus is that cable audience data also cannot be measured in the same diaries that presently are used to gather broadcast audience information because:

- By and large there is not adequate space for listing all of cable's multiple channels.

- There is a general confusion on the part of the viewer as to identification of the different cable channels.
- Just as it has long been felt that diaries tend to penalize the lower-rated, occasionally viewed broadcast programs, there would be an even greater problem with cable's small audience levels.

To a certain extent, the problems in measuring cable audiences may be similar to those involved in radio broadcasting measurement. Radio broadcasters have long been concerned about how to measure fractionalized listening to 30, 40, or more stations in a market. With

TABLE 8-5 Multiple Channels to Measure

Channel	Programming	Channel	Programming
2A	Chicago CBS	3B	Library Access
3A	Promos	4B	Educational Access
4A	Milwaukee NBC	5B	Local Government
5A	Chicago NBC	6B	Consumer News
6A	Milwaukee CBS	7B	Local Origination
7A	Chicago ABC	8B	Government Access
8A	News Features	9B	Public Access
9A	Chicago (WGN)	10B	Time/Weather
10A	Milwaukee Educational	11B	Bulletin Board
12A	Milwaukee ABC	12B	Swap & Shop
17A	Atlanta (WTBS)	13B	Government Access
18A	Milwaukee Independent	17B	Showtime
20A	ESPN	18B	HBO
21A	WHA (Educational)	20B	The Movie Channel
22A	Job Bank	21B	Cinemax
23A	C-SPAN	22B	HTN
24A	Chicago Independent	23B	Bravo
27A	Financial News	24B	Galavision
28A	New York (WOR)	27B	Swap & Shop
30A	CBS Cable	28B	Nickelodeon
32A	Chicago Independent	30B	CBN
33A	National Jewish TV	32B	USA
36A	Milwaukee Educational	33B	CNN
37A	MSN	34B	National/Internat'l News
38A	PTL	35B	Sportswire
39A	NCN	36B	State/Local News
40A	Trinity	37B	AETN
41A	BET	39B	SIN
2B	Channel Guide	41B	MTV

cable, these numbers of channels will be commonplace. For example, Table 8-5 shows Viacom's Greendale, Wisconsin, system back in March 1982 when 60 out of a possible 82 channels were being programmed.

FIGURE 8-1 Correct Base for Cable Audience Measurement

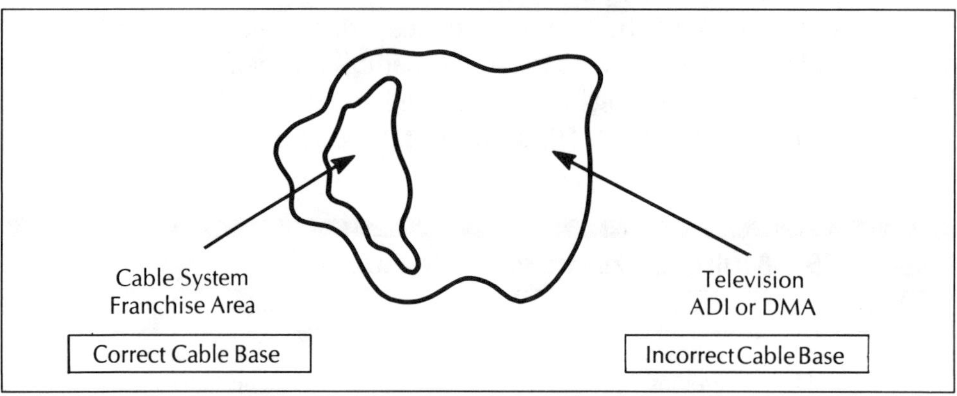

Researching the Research

The Cable Audience Methodology Study (CAMS) was aimed specifically at researching cable research techniques. The study's objective was not to develop a rating system per se, but to spearhead the evaluation of various methodologies for measuring local cable audiences and to determine which methodologies might represent the most viable, reliable, and cost-effective method or methods for obtaining viewing data for programming and advertising purposes. Behind the study was a committee made up of representatives from major firms in both the cable and advertising fields. This effort was the first time that the buyers and sellers of advertising time cooperated in sponsoring such a project.

The A.C. Nielsen Company was selected to conduct the Cable Audience Methodology Study under the supervision of the Industry Research Standards Committee, jointly sponsored by the NCTA and the Cabletelevision Advertising Bureau.

Seven separate tests were conducted to gather information on household viewing and to determine personal cable usage. Diary and telephone collection techniques were evaluated, and both daypart and quarter-hour viewing levels were tested. All of these data were validated against a series of telephone coincidental surveys, which, for many years, have been accepted as the most reliable way to establish accurate audience levels.

When results of the study were released in 1982, they confirmed what had long been felt—that the standard television diary tends to understate cable viewing by very large amounts in some time periods. It further pinpointed specific problems involved in economically and accurately measuring both broadcast and cable audiences using the same techniques. The results of CAMS will help form a basis for determining "How does cable solve its research dilemma?"

The Creative Cable Research Kit

The needs for research vary greatly depending on who is using it and what they are using it for. For an advertiser making a direct response offer, research is built in. It's the response delivered.

For most advertisers, however, this is not the case. An advertiser who spreads advertising messages throughout a cable channel on various days and at various hours may well find that a cumulative audience measure of those people who view these time periods at one time or another may be adequate. If, however, that advertiser selects specific programs at different times—where there is not a continuity of viewing—there would be a greater need for individual program rating data. For some advertisers (for instance, local advertisers who are presently using local shopping newspapers and have that kind of information now available to them), it may suffice to have an adequate measure of the distribution of the cable homes and the characteristics of the subscribers.

They say that necessity is the mother of invention, and, in lieu of regularly reported audience data, many advertisers, agencies, cable networks, and system operators have been extremely imaginative. An advertiser can even put together a "Creative Cable Research Kit" by investigating the variety of efforts thus far. Some examples follow.

The "Daytime" Woman

Before the Hearst/ABC "Daytime" service went on the air in March 1982, their research director sought to identify its potential audience. The objective was to define who the "Daytime" woman was and to take a look at her demographics, psychographics, activities, and consumer habits.

Hearst/ABC examined "Daytime's" Projected Viewing Areas—those municipalities that were carrying the SATCOM I Transponder 22 signal from 1:00 to 5:00 P.M. EST (the time period to be occupied by "Daytime"). The PVA's were identified by five-digit zip codes that were matched against the zip codes on the Simmons Market Research Bureau's 1980 data file.

The characteristics of a representative sample of women who were cable subscribers living in the PVA's were then compared to a sample of all adult women nationally. Hearst/ABC used this comparison to show how the "Daytime" woman compared favorably in demographics, psychographics, and consumer behavior with all women. (See Table 8-6.)

Research like this must be examined very carefully. It assumed that the "Daytime" viewing audience would actually parallel that of women living in areas that receive the "Daytime" signal. Such may not in fact be the case. Nevertheless, it is an imaginative effort to provide projected audience characteristic data when none are available, and it can be a very effective sales tool from the standpoint of the cable network and cable system.

CBS Cable and the Upscale Viewer

In a project similar to the "Daytime" study, CBS Cable wanted to show the upscale nature of their audience in November 1981, at a time when audience information on it was not yet available. As with "Daytime," CBS Cable used Simmons Market Research Bureau data. In this case, they analyzed the demographics, product purchase habits, and media usage of households that reported viewing cultural and performing arts television programming in the Simmons National Study of Media and Markets. CBS Cable stipulated to Simmons that all households used in this analysis had to be located within the coverage area of cable systems carrying their new cultural network. The results showed that these viewers were affluent, professionally successful, and well-educated and that they were consumers of expensive products and services and regular readers of upscale print media. For example, Table 8-7 shows their reading preferences compared to the national average.

Once again, the assumption was made that the CBS Cable viewing audience would have the same characteristics as those people in the Simmons national diary sample who were located in the coverage area of cable systems carrying CBS Cable and who viewed

TABLE 8-6 Characteristics of "Daytime" Female Cable Subscribers Compared with All Adult Women

	"Daytime" Percent	National Percent	"Daytime" Index
"Daytime" Household:			
Household size 3+	64.0	54.0	118
Number of 18+ adults	30.0	26.0	115
Presence of children	51.0	44.0	116
Married	66.0	61.0	108
Married 10 years+	75.0	70.0	107
Live in top 25 ADI's	61.5	53.8	114
"Daytime" Woman:			
25-49	47.6	44.8	106
Attended college	32.3	27.7	116

Source: Simmons Market Research Bureau.

cultural and performing arts programming. In this case, the assumption was probably fairly valid, and it does provide a cable network and potential advertiser with some relevant research data where none would otherwise exist. Unfortunately, research could not keep CBS Cable alive!

Who Watches the "Home Shopping Show"?

In some cases, research can be generated by a follow-up questionnaire to people who respond to direct response offers. The Modern Satellite Network's "Home Shopping Show" did this in a survey conducted by the A.C. Nielsen Company. A questionnaire was mailed to 1,000 people who had used a toll-free telephone number to respond to offers made by advertisers on the "Home Shopping Show."

From 355 returned and completed questionnaires, the following results were tabulated:

- Customers had higher than average household incomes—40 percent over $30,000.
- Respondents were frequent purchasers of appliances, with 26 percent having purchased a television set in the previous 12 months.
- More than half the respondents were women, with 28 percent in the 18-34 age group and 25 percent in the 35-54 age group; 26 percent of respondents were men 18-54.

- Respondents were heavy users of coupons, with 91 percent coming from households where newspaper or store coupons had been used to purchase a product in the previous 30 days.
- Three-fourths came from households in which a member had ordered a product from a catalog in the previous year; 49 percent made more than three such orders a year.
- When asked their reasons for watching the show, 95 percent cited the show's informational value; 52 percent said they watched the show twice a month or more.
- Nearly a third (32 percent) of viewers said "The Home Shopping Show" influenced them to purchase a particular product.

Since the "Home Shopping Show" is telecast during the day time, some advertisers questioned whether it reached an affluent audience. The purpose of this research by the Modern Satellite Network was to show that the "Home Shopping Show" had a very active, involved audience—that is was not simply a passive daytime television viewing audience. Again, a research technique was devised to produce qualitative data where no regular survey information existed.

Developing a Local System's Profile

At the local level, *Cable Plus,* a weekly cable and network TV listing publication in the Dallas/Fort Worth area conducted a study defining

TABLE 8-7 Reading Preferences of Audience in CBS Cable Area

Publication	Percent Greater than U.S. Average
Business Week	+38
Forbes	+32
Fortune	+87
Scientific American	+45
U.S. News	+49
Wall Street Journal	+43
Any Epicurean Publication	+49
Omni	+84
New Yorker	+83

Source: SMRB Special Analysis (November 1981)

the magazine's audience for advertisers. Of 500 questionnaires delivered to subscribers, 252 were returned. *Cable Plus* used the survey results not only to reflect the characteristics of their reading audience, but also to present a profile of the market's cable audience. They assumed that this was a logical assumption to make since its readers all subscribed to cable. The results showed an affluent audience that devoted approximately 23 percent of its viewing time to the cable networks (46 percent to the premium cable channels) and owned a significant number of video cassette recorders and video games.

The *Cable Plus* study is one approach to gathering information on a local market basis. Another way is for an individual cable system to include with its monthly bill a questionnaire on viewer interests and behavior.

Polling via Telephone

No one would question the fact that if television programmers knew, before putting a show on the air, whether people would want to watch it or not, they would stand a better chance of getting a high rating. WTBS and Bristol-Myers began doing this in February 1982 when they allowed viewers the opportunity to select the Bristol-Myers Theater motion picture via AT&T's "Dial-It Telephone Polling System."

During the January 4 Bristol-Myers movie, viewers nationwide were told that they could call a designated telephone number to vote their preference on a given film for February. There were six possible movies corresponding to six different numbers viewers could dial to indicate their preferences. Results were tabulated instantly by AT&T and announced by WTBS personality, Bill Tush, immediately upon the conclusion of the January 4 film. "Paint Your Wagon," the movie that received the most votes, was then scheduled for February 1. (Overall results are shown in Table 8-8.)

Since the first Bristol-Myers Theater "Dial-It" survey in January 1982, viewer response has grown considerably. During the June 6, 1983 showing of "Spencer's Mountain," a total of 28,436 viewers voted on their choices for July and August.

Recipes, Cookbooks, and Research

Analysis of the distribution of response to offers also can provide some guidance as to the distribution of the viewing audience. This is actually a very old technique. Back in the early days of radio, before the development of audience measurement services, Procter & Gamble offered flower seeds on one of their radio shows. By analyzing where the responses came from, they had some idea of where the radio show's listeners were.

Kraft used a response mechanism to determine if—in lieu of traditional rating data—there was anyone out there watching. They didn't believe the people who said, "If you can measure something in a rating report, it exists and if you can't, it doesn't exist." Kraft offered a free recipe book to anyone who wrote to WTBS and the Cable News Network. The cable advertising schedules were small and ran only three weeks. Kraft had no solid rating data when the buy was made and didn't know whether the offer would pull 100 or 1,000 requests.

The main point of the test was to see if, lacking this solid audience data, there was anyone out there to jiggle the meter. In total, over 10,000 requests came in. Considering that the offer was a modest one and viewers had to be alert to catch it, this response was outstanding. Kraft found that an active, alert audience did exist for cable even though they couldn't measure it in traditional terms. And to gather even further data, they surveyed a sample of the respondents with a simple postcard questionnaire asking about their product usage. (See Figure 8-2.)

TABLE 8-8 Results of "Dial-It" Telephone Poll

Movie	Response
Paint Your Wagon	1,329
Damn Yankees	804
Flower Drum Song	747
Angel in My Pocket	627
Greatest Show on Earth	577
Night and Day	480
	4,654

Ongoing Research in Madison

Madison Cable Network in Madison, Wisconsin, has developed an ongoing research program to show advertising prospects the impact of cable in the market. To demonstrate how network audience ero-

FIGURE 8-2 Kraft Cable Questionnaire

Would you answer a few questions for us?
and please drop this card in the mail.

1 Often food companies such as Kraft feature recipes which specify their brand(s). The first time you try one of these recipes how likely are you to use the specified brand?
( ) **Almost always** ( ) **Usually** ( ) **Sometimes** ( ) **Never**

2 Now, when you use the recipe a second or third time, are you more likely to continue to use the specified brand or are you more likely to make a substitution?
( ) **Use specified brand** ( ) **Make a substitution**

3 Please look through the recipes in the Kraft booklet you just received. On the lines below, list any Kraft products you regularly buy and that you would use in order to make one or more of these recipes.

KRAFT PRODUCTS: ______________________

____

____

4 On the lines below, please list any Kraft products that you do *not* usually buy, but *would* buy in order to make one or more of these recipes.

KRAFT PRODUCTS: ______________________

____

Thank you so much

sion becomes translated at the local level, Arbitron conducted three viewership studies in cable households for MCN in 1983. In the September 1983 survey, three network prime-time shares totaled only 49 percent. (See Table 8-9.)

TABLE 8-9 Madison Audience Shares

Pay Cable (HBO, SHO)	20%
NBC	20
Basic Cable	18
CBS	18
Cable Superstations (WTBS, WGN)	11
ABC	11
PBS	2

Source: Arbitron Telephone Coincidental (September 1983).

Then to detail the upscale demographics of the cable subscribers, the Midwest Survey Research Institute conducted a qualitative survey of Madison's Complete Channel TV (CCTV) subscribers and found that:

- 59 percent were 18-44.
- 71 percent were married.
- 70 percent owned their own home.
- 37 percent had a pretax income of more than $35,000.
- 80 percent attended or graduated from college.
- 65 percent reported watching as much or more cable than network.

Madison Cable Network uses these studies in combination to demonstrate the values of cable as a local advertising medium.

Leo Burnett and MTV

According to a Leo Burnett Company study of advertising-supported cable networks, MTV had attracted a substantial audience only six weeks after its debut in August 1981. A telephone survey by Leo Burnett's Media Research staff of cable subscribers to a single system in McHenry, Illinois, showed that familiarity and viewing levels for MTV: Music Television equaled or surpassed that of older services such as CNN, ESPN, and the USA Network. While this was a relatively small survey in only one county, later research by MTV: Music Television substantiated the results.

Viewing in the Valley

Valley Cable in West San Fernando Valley conducted a survey to help in its programming decisions. Among the questions Valley Cable asked of their subscribers were several on viewing to the ad-supported networks for which Valley Cable sells time. The results showed that MTV had more viewers that watched 2+ hours a day than the next three networks combined. (See Table 8-10.)

Though the sample was small and from only one cable system, the research reflected Nielsen meter findings showing that MTV had the highest ratings of any basic cable network. The combination of such a *local* system survey with *national* meter data can be used by the system as a valuable tool in selling local cable availabilities.

TABLE 8-10 Valley Cable Viewing

	Watch Cable Network 2+ Hours Per Day
MTV	30%
CNN	10
ESPN	9
USA	7
Total Respondents	100% (191)

Source: Telephone Survey of Valley Cable Subscribers, conducted by Valley Cable TV, West San Fernando Valley, August 1983

A Cable Subscriber Study

A cable system operator knows that to maximize profits, it is necessary to achieve the highest possible subscriber base with the lowest possible monthly "churn" (or turnover). To do so, cable operators must be ever mindful of the needs of the community they serve and offer services that the community wants. It is important that systems constantly assess the interests of their communities in order to know which cable services to add and which to drop.

Gabe Samuels of J. Walter Thompson USA developed a format for a simple subscriber study that local systems can use periodically to assess viewer preferences and service satisfaction. This survey can be adapted to almost any system with questionnaires distributed in the monthly statements or as special mailings. (See Figure 8-3.)

The Old Spice Quiz

In an earlier chapter, I mentioned how Old Spice created a sports trivia contest that centered on events and stars in baseball, basketball, boxing, football, and hockey. Commercials ran in a variety of ESPN sports with each program's responses identified by a different code. Old Spice was in effect developing its own cable rating system before rating data were generally available on ESPN.

Multiple Research Opportunities with Qube

While the Warner Amex Qube system is generally thought of as providing a new addition to programming in the form of two-way feedback, the technology has many research applications. Subscribers

FIGURE 8-3 Cable System Subscriber Questionnaire

(Date)

(NAME OF TOWN) CABLE TELEVISION
ADVISORY COMMITTEE

Dear (Name of Town) Resident:

Your cable TV committee has been appointed to advise the *(City Council, Town Board, etc.)* as to the performance of *(Cable System).* The attached questionnaire is our official request for information from residents of this town.

Since it would have been impractical for us to question every single household in the town or tabulate the answers—we have randomly selected a representative sample of the total population.

We need a 100% response to insure that the results of this survey are representative. It is, therefore, necessary that *everybody* reply.

For everybody's convenience, we enclose a self-addressed, post-paid envelope in which you can mail your questionnaires back to us.

If you are unable or unwilling for *any reason* to respond to this questionnaire, please call us at *(Phone Number).* (Leave a message for the Cable Committee.)

Thank you for your cooperation.

PLEASE RETURN WITHIN 7 DAYS

QUESTIONNAIRE

1. Are you currently a cable TV subscriber? Yes ☐ No ☐
   If you answered yes, please skip to question 5.
2. Have you previously subscribed to the cable service right here in *(Name of Town)*? Yes ☐ No ☐
   If you answered yes, please skip to question 4.
3. Is cable TV service available to you? Yes ☐ No ☐ Don't know ☐
   To be answered only by those who are *not* current or previous subscribers.
   If you have answered this question, there's no need to answer any others. Just return the form to us in the enclosed envelope.
4. Why did you cancel the cable service?______________________________

   If you have answered this question, there's no need to answer any others.

5. What is your monthly payment to *(Cable System)?* $__________

6. Date cable TV was installed in your home? __________

7. How much did you pay for the original installation? $__________

8. How many sets in your home are hooked up? __________

9. Have you had any problem of the following kinds?
   1. Installation Yes ☐ No ☐
   2. Quality of reception Yes ☐ No ☐
   3. Service calls Yes ☐ No ☐
   4. Billing Yes ☐ No ☐
   5. Program quality Yes ☐ No ☐
   6. Other Yes ☐ No ☐

   If any of the "yes" boxes were checked, please explain. Use the back of this page if necessary.

   ____
   ____
   ____
   ____

10. Did you report the problem or problems to *(Cable System)?* Please supply details. Use the back of this page if necessary.

   ____
   ____
   ____
   ____
   ____

11. The following is a list of services (channels) now available on your cable system. Please indicate the *degree* to which you would like to continue having this service available to you. ("5" is the highest interest, "1" is the lowest.)

	Check If You Now Receive	Low 1	2	3	4	High 5
(List and Description of Services)	( )	( )	( )	( )	( )	( )

12. The following is a list of services (channels available on cable systems *elsewhere.* Please indicate the degree to which you would be interested in having these available to you. ("5" is the highest interest, "1" is the lowest.)

	Low 1	2	3	4	High 5
(List and Description of Services)	( )	( )	( )	( )	( )

have consoles with at least 5 (in several cases 10) response buttons. This ability to respond to interactive questions allows Qube to deliver:

- Market Research. Cable can be used to demonstrate and test product concepts, copy, programs, packaging and other visual material. The audience for this research would be either actual viewers or a prerecruited sample of persons with selected characteristics.
- Media Research. Qube's electronic polling capacity can provide ratings and shares, audience cumes and frequency and channel-switching patterns.
- Program Research. Qube can provide almost immediate viewer feedback on programs and program concepts.
- Public Opinion Research. The interactive capability of Qube can measure opinions on political and social issues among actual program viewers or a prerecruited target group or panel.

While Waiting for the Ratings to Come In!

In early 1984, the Financial News Network was waiting for the A.C. Nielsen Co. to provide metered audience data to them. In the meantime, it was important to constantly re-affirm to potential advertisers that they were reaching an upscale audience with very little waste. To do this, they ordered a special FNN Demographic Telephone Survey from Nielsen profiling people who respond to PI offers on FNN. It pointed out that the FNN PI responder was:

- Male
- 25-64 years old
- Highly educated (Post Graduate)
- A home owner
- A new car buyer
- A 35mm camera, VCR, PC owner
- A multi credit card holder (three average)

- Upper management or business owner
- A frequent airline traveler
- In the upper income bracket ($50-$60,000)

In interpreting the results of this study, it is important to recognize that this survey was profiling only those FNN viewers who responded to a direct response offer and not all FNN viewers.

FIGURE 8-4 Advertising Impact Scoreboard

	✓'s	Totals
Cable		
Radio		
Television		
Outdoor		
Direct Mail		
Daily Newspaper		
Suburban Newspaper		
Word-of-Mouth		

The "Advertising Impact Scoreboard"

If a cable system operator is interested in actively soliciting local advertising in a community, he or she will find that a common response from the merchants is, "But how will I know if anyone sees my advertising?" As a promotion idea, the cable system might provide its advertisers with an "Advertising Impact Scorecard" like the one shown in Figure 8-4.

When a customer calls or stops into the store, the owner asks, "How did you find out about our sale (or our special promotion)?" Or, "Where have you heard about us lately?" He checks off each response on the "Advertising Impact Scorecard" and can see at a glance where the advertising response is coming from. *If this seems risky, the cable system operator must remember that, in the long run, cable will succeed as a local advertising medium only if it produces results.* It is up to each cable system to work hand in hand with local advertisers to assure good results by helping them in the execution, promotion, and evaluation of their advertising.

Cumulative Audience: A Helpful Tool

Cumulative audience data can provide a clue as to the broad potential of a cable service to generate audience based on viewing over an extended period of time. The Arbitron National Network Cable Report provided such data in their first viewership study of eight pay and basic satellite programming services. Table 8-11 shows the weekly cume ratings from this November 1981 study.

TABLE 8-11 Arbitron Weekly Cumulative Adult Ratings

	Percent U.S. Adults Able to Receive		
	Adults 18+	Men 18+	Women 18+
CNN	57	61	53
CBN	30	26	33
ESPN	48	61	36
HBO	91	92	90
SHOW	87	89	86
SPN	35	34	35
USA	42	48	36
WTBS	68	72	64

Source: Arbitron, November 1981.

VideoProbeIndex was a continuing tracking study of the new electronic media. It showed cumulative audience levels reported for 15 services. (See Table 8-12.)

By comparing a satellite service's 1-day cume with a 7-day cume and then with an even broader 30-day cume, an advertiser could determine what percent of the potential audience (on a monthly basis) could be delivered in a shorter period such as a week or day. For example, VideoProbeIndex reported that, in the course of 30 days, WTBS was viewed in 52 percent of its coverage area's households. It reached over 90 percent of this audience in the course of a given week—63 percent of it on a single day. (See Table 8-13.)

TABLE 8-12 VideoProbeIndex Cumulative Household Ratings

	1-Day		7-Day		30-Day	
	U.S. Cable (Percent)	Coverage Area (Percent)	U.S. Cable (Percent)	Coverage Area (Percent)	U.S. Cable (Percent)	Coverage Area (Percent)
CNN	24.1	53.9	35.5	79.3	40.0	89.4
USA	6.7	16.9	13.1	33.1	16.0	40.4
ESPN	19.9	35.1	35.4	62.4	40.9	72.1
SPN	2.4	14.7	5.4	33.1	7.7	47.3
CBN	5.7	9.5	10.9	18.2	15.1	25.2
MSN	0.4	2.3	0.9	5.3	1.3	7.6
Nickelodeon	3.2	12.4	7.3	28.3	10.3	39.9
ARTS	0.9	3.5	1.8	7.0	3.9	11.6
MTV	3.6	34.6	5.8	55.7	6.9	66.2
C-SPAN	0.9	NA	2.0	NA	3.7	NA
BET	0.2	0.6	0.6	1.6	1.3	3.5
SIN	0.3	NA	0.6	NA	0.9	NA
WTBS	27.1	33.3	38.7	47.5	42.7	52.4
WGN	15.1	56.8	22.8	85.7	25.6	96.2
WOR	14.1	NA	24.4	NA	28.0	NA

Note: VideoProbeIndex developed cumulative ratings based on a total U.S. sample and projected these to each network's coverage area based upon December 31, 1981 estimates of each network's household base.

Source: VideoProbeIndex (August 1981)

Many advertisers buying schedules on a given satellite network want to know the *specific* cumulative audience—or reach of their *specific* buy.

At this early stage of cable research, the estimates of the four-week reach for various level schedules of cable-originated programming shown in Table 8-14 were produced. Of course, it is necessary to know the average ratings for the schedules involved and

TABLE 8-13 WTBS Cumulative Household Ratings

	Coverage Area Percent Households
1-Day Cume	33.3
7-Day Cume	47.5
30-Day Cume	52.4

Source: VideoProbeIndex (August 1981)

TABLE 8-14 Four-Week Reach for One Cable Satellite Service (Assuming One Commercial per Program)

	Net Reach (Percent)	
4-Week Target GRP Levels	At a 1 Rating or Less	At a 2 Rating
20	8	9
25	9	11
50	14	16
75	18	20
100	21	24
125	25	28
150	28	31
175	30	33
200	32	35
225	34	38
250	35	40

a good estimate is that most existing cable satellite services average a one rating or less. The exception is WTBS, which averages over a 2 rating in its coverage area. (See Table 8-15.)

TABLE 8-15 The Average Minute Audience of Nine Cable Networks

	Coverage Area Rating (Percent)
WTBS	2.5
MTV	1.2
USA	1.1
CBN	1.0
ESPN	0.9
CNN	0.8
CNN Headline News	0.5
The Nashville Network	0.5
The Weather Channel	0.4
Lifetime	0.3

Source: Estimates based on 1984 Nielsen Reports (7am-1am)

Be a Cable Research Explorer

Over 20 years ago in January 1964, Arthur C. Nielsen, Sr., announced that Nielsen was terminating its network radio rating service. Mr. Nielsen said, "To my great disappointment, events beyond our control have changed the character of network radio. The average radio program rating has declined due to the combined effect of television and the tremendous increase in the number of radio stations competing for the available audience—approximately four-fold since 1946! Furthermore, radio listening, formerly limited largely to plug-in receivers, is now divided three ways—plug-in, battery portable, and automobile. Thus, there are now three measurement tasks instead of one!"

By changing only one or two words or key phrases, that statement could apply to cable today. It would indicate that the job of measurement is just too great to be accomplished. But to say that would be like the ostrich who sticks his head in the sand when problems occur. And cable is a place for explorers—not ostriches!

CHAPTER 9

Cable at the Local Level

"All business is local!"

This statement has long been accepted as fact by marketers of goods and services throughout the country.

For no matter how much attention is focused on the billions of dollars spent each year on national advertising, the final decision to buy or not to buy anything from a new brand of canned peas to a new car is made by an individual consumer in his or her own community.

The Local Impact of Cable

In large cities, cable will not so much provide advertisers with a new medium as it will let them deliver a unique creative treatment that existing broadcast television does not provide. This includes in-depth personal selling, advertising tailored to a specific environment, and affordable program sponsorships as existed in the early days of television.

In the small towns and suburban areas, however, cable will be a *brand new media form.* Television has generally not been practical either because of its waste coverage outside of these communities or because of its price tag. Cable, however, offers low-cost, informational, service-oriented advertising that is tailored to the consumers concentrated within these small geographic areas.

Bill Sanders, 1984, *The Milwaukee Journal*

Each week, new towns and cities across the country are taking their first steps in the new electronic era by awarding cable franchises and beginning the process of becoming wired up. The impact of cable in each of these communities will take place at two levels—among the viewers of cable and among the advertisers who will use cable as a new selling medium. From the viewers' standpoint, a new satellite menu of news, sports, entertainment, information, and education will be further supplemented by a wide array of local community programming and neighborhood events. From an advertising standpoint, marketers will be able simultaneously to develop ways of using cable more effectively to promote and sell their products and services in these communities.

The promise of cable at the local level is to open up the video medium to advertisers in a new, affordable manner. With its dozens of channels, cable will also create program opportunities for advertisers too limited in audience appeal to be attractive to the larger broadcast television stations. The result is that local marketers will find new ways to attract the customers most interested in their products and services.

The opportunities are endless — limited only by the imagination.

What's Happening Out There?

A Quick Survey of 12 Ideas

Let's take a quick trip around the country and examine some ways in which local cable advertising has been used:

- A cable system in Moline, Illinois, leases a channel to a local real estate company. Their show is 24 hours a day of photos and descriptions of property for sale.
- In Naples, Florida, Palmer Cablevision creates local advertiser-related programs that include a garden shop, a home exchange, a money show, and a boating feature.
- In the suburbs of Boston and in New Bedford-Fall River, Massachusetts, four cable systems program 7 to 15 hours a week of local programming in which they sell advertising.
- Viacom Cablevision of Long Island solicits business advertising on its data channel. Responses were so good that they developed a special restaurant listing and car dealer listing inserts.
- Just prior to Christmas, Joyce Cable of Romeoville, Illinois, introduced "Christmas Greeting Cards" for local businessmen and politicians to air holiday greetings via cable.
- In Athens, Georgia, Liberty Cable sent a production crew door-to-door among the business community offering ten 15-second Christmas commercials to run through the holiday period at a very attractive one-time price. Merchants featured one product or service and made a special "cable offer." The spots were then combined into two-minute commercials and aired in the local availability slots of Cable News Network. The merchants reported a large increase in

store traffic, and Liberty Cable learned how to put commercials together from raw footage.

- Continental Cablevision of Lansing, Michigan, goes after local businesses to demonstrate their wares. A salesperson does a half hour on how to shop for a stereo. A local lumber dealer demonstrates how to build and hang a suspended ceiling.

- Albuquerque Cablevision couples local commercials run on satellite services, such as CNN and ESPN, with advertising inserts in their monthly subscriber billing statements. An advertiser's coupon or ad is inserted directly into the envelope with the subscriber's monthly statement. In other markets, cable operators are similarly finding it is quite effective to offer local clients a package combining cable spots with cable guide ads and bill stuffers.

- In Grand Rapids, Michigan, Computerland sponsored a 28-minute infomercial produced by Apple Computers. This infomercial was promoted via newspaper ads and took advantage of cable's potential to provide a detailed picture of a product sold by this computer retailer.

- CPI of Louisville carries a real estate listing and home improvement channel. During the day, there are 10-second clips with audio of homes for sale, plus longer messages for apartment and condominium complexes. In the evening, these listings are intermingled with home care programming. One of CPI's first efforts at selling advertising was its "wholesaling" of all local availabilities on the Satellite Program Network to Smith Furniture and Supply Store. Smith ran different messages to coincide with the appeal of each SPN program—women, financial, etc.—and used co-op money from General Electric for a major part of the buy. CPI promoted SPN in Louisville as the "Smith Program Network" and incorporated the store's identification into its promotion.

- In New York, Manhattan Cable offers cable classifieds to sell cameras, cars, stereos, or promote instructional or specialized services, such as acting, or singing lessons, catering or counseling. They even provide a "cablegram"—a video telegram for personal messages such as birthdays, anniversaries, and graduations.

- In Chapel Hill, North Carolina, Village Cable features a Home Shopping Channel which runs from 7 P.M. to midnight.

Profiting from Local Origination

At the local level, cable systems are looking for imaginative ways to turn their local-origination channels into advertising revenue generating businesses. One example is "The TV Automart," a video version of newspaper classified advertising that Group W Cable developed for its systems in the Los Angeles area. The hour-long show allows private-party auto sellers to do their own 90-second "infomercials" showing and describing their cars. An answering service lets potential buyers call and place their bids on the advertised autos. In addition to private-party selling, "The TV Automart" offers time for businesses in the automotive aftermarket to demonstrate their products or services.

Another attempt to turn local-origination channels into commercial ventures is "Hollywood Weekly." It is a video movie guide with star interviews and show business news items, in which movie companies pay to run their film promotions and listings where they are playing.

And in Phoenix, R/G Cable, a division of Phoenix Newspapers, introduced the Home Channel as 24 hours of videotaped informational advertising about homes and services associated with home ownership.

Each of these ventures has one common purpose—the development of new advertising revenue streams from local-origination channels.

At-Home Browsing and Shopping in Greenbay

In Greenbay, Wisconsin, the Home Video Shopper operates 24-hours a day, seven days a week providing cable viewers with at-home browsing and shopping. Every commercial message runs six times a day, and all advertising is grouped in a related category. Automobiles, auto repair, and tires are together, as are homes, interior decorators, paint, etc. The HVS producer-director works with the advertiser in developing messages from 30 seconds to 60 minutes in length. Just like a local newspaper creates special advertising sections, so does the Home Video Shopper. A Christmas Gift Guide meets advertisers' holiday advertising needs in November-December. A Dining and Entertainment Guide is designed for local restaurants, nightclubs, lounges and ballrooms.

"Auto Buyer '84" was devoted to the arrival of 1984 automobiles and trucks. It featured both videotape presentations of the major auto and truck manufacturers and advertising by dealers, auto accessory stores and repair services. In addition to reaching the 25,000 cable subscribers in Greenbay, "Auto Buyer '84" was fed to an additional 40,000 cable homes in northeastern Wisconsin and Upper Michigan.

Turning a Live Event into a Cable Event

To utilize local cable advertising most effectively, a business must consider what the medium offers that cannot be found in any other medium. In some cases, the answer can be very obvious, yet exciting.

Here is an example. A major store in the Southwest sent out invitations to its customers to join them in a series of mini-seminars on health and beauty care. *On Monday*, there was a seminar on skin care and make-up instruction for the career woman. *On Tuesday*, the store had a program on how to update and build a wardrobe. *On Wednesday*, there was a discussion of hair care and new hairstyles. *On Thursday*, the authors of a new book on aerobics demonstrated basic exercises and dance—with a presentation of summer swimwear.

Sometimes the most obvious opportunity is the most elusive, and in this case, it was. Only after the four-day program was completed did the director of advertising realize that the promotion effort could have been extended by having the local cable system tape the different events and present them as a video promotion accompanied by the latest in summer fasions and beauty care. The store could have expanded the number of women participating in the event many times.

Lakes Cablevision—A Success Story

Lakes Cablevision, a 36-channel system that serves over 10,000 households in McHenry and Lake County, Illinois (about an hour's drive north of Chicago) has taken a very aggressive stance in developing local programming and selling advertising.

Marketers have their choice of running local spots on the major satellite networks or advertising on a wider array of local programming that includes special features such as high school sporting events, band concerts, plays, local news interviews, and celebrity and political interviews.

Originating 40 hours of local programming a week, Lakes Cablevision encourages advertisers to sponsor such special features as:

- "Primary Source"—a look at the world of nutrition and foods and the people who grow and process them, with homemaking hints including recipes, decorating ideas, and crafts.
- "New and Now"—interviews with professionals in the field of beauty and personal care, focusing on hair care, fashion, nutrition, make-up, skin care, and exercise.
- "Careers"—help for the young person in choosing a career and getting a job.
- "People Issues"—a feature program that covers diverse topics of concern from alcohol and drug abuse to aging and child care.
- "Chamber Spectrum"—focusing on Chamber of Commerce activities going on around the area and featuring guests to talk about their businesses and the services they offer the community.
- "Your Chance to Live"—what to do when an emergency strikes.
- "Hook, Line and Sinker"—aimed at the fisherman with tips on when and where the fish are being caught and how to catch them.

The Selling of Local Cable Advertising

The preceding examples of advertising at the local level have been the exception rather than the rule. Today, local advertising availabilities are offered in a variety of satellite services such as:

Cable News Network

CNN Headline News

ESPN

Lifetime

MTV: Music Television

The Nashville Network

USA Cable Network

Satellite Program Network

The Weather Channel

Most systems, however, are not even selling these local opportunities much less developing unique advertising concepts. Rather than invest in the sales resources and equipment necessary to succeed in the advertising business, most local systems have directed their efforts toward acquiring subscribers and improving services.

Through the combined creative efforts of a system's customer sales and advertising staffs, however, a cable operator could develop programs to actually increase the viewers' cable satisfaction through the advertising stream. For example:

1 Viewers would be offered price incentives on the purchase of home electronic equipment to focus on "The Cable Home Entertainment Center." These would include TV sets for multiset hookups, VCRs, special tape rentals, and stereo receivers. In exchange for these special price incentives, participating local electronics dealers would be given special discounts on advertising rates.

2 Nothing goes together like popcorn and movies, and popcorn poppers from local appliance dealers could be offered as premiums for signing up for movie channels.

3 Cable contests—mini-quizzes on local channels with questions related to cable programs and the community—would have prizes from local advertisers.

The objective is to offer the subscribers a little something extra and to create new ways of generating advertising revenue from local merchants.

Another creative approach to local cable marketing that ties in with advertising is a Cablefest in which a cable system demonstrates to a local community the wide range of program options it offers. Viacom Cablevision of Wisconsin planned a Cablefest at a major Milwaukee area shopping center to inform subscribers and create awareness among those who did not yet have cable. During the Cablefest, The "Weatherwear Fashion Show" afforded a showcase for many of the mall merchants and was a natural tie-in promotion of The Weather Channel. It featured fashion as it related to weather conditions, including ski apparel, rainwear, and cold weather clothing.

Weather Channel participants included the affiliate manager, who served as one of the models, and an on-camera meteorologist as Master of Ceremonies. The fashion show concept was developed to highlight The Weather Channel's many special features that are designed to assist the viewer in planning his everyday activities. For instance, ski fashions were presented, which tied in to the "Skiers' Forecast."

By participating in such local cable marketing efforts, national satellite networks have an excellent opportunity to increase their own awareness in a market and help build both subscribers and viewership. At the same time, local merchants can be made more aware of cable and become better advertising prospects.

Getting Started

Local systems obviously need help in evaluating the advertising potential of their markets and in setting up effective sales programs. Assistance is coming from the Cabletelevision Advertising Bureau in the form of cable advertising profiles, management reports, seminars, and consultations. Most cable operators who have been successful in selling advertising time agree that any cable system can succeed if it researches its market and hires a professional to bring in the advertisers.

It also helps to use creativity and imagination.

To introduce cable advertising in Grand Rapids, Michigan, GE Cablevision staged a one-night workshop highlighted by Las Vegas games of chance. Local advertisers played "no-loser" roulette, poker, and blackjack in which they won coupons good for free spots on the system's satellite advertising networks—CNN, ESPN, SPN and USA. The objective was for advertisers to use the coupons and, hopefully, buy additional spots. This goal was accomplished.

In another unusual kick-off of a local advertising effort, Daniels and Associates sold half of the CNN and ESPN availabilities on its Ann Arbor, Michigan, system to one local advertising agency at $1 per spot. The philosophy was that it is better to have the advertising time filled then go unsold. Furthermore, when potential local advertisers saw that their peers were using local cable system, they were encouraged to follow suit.

Madison Cable Network, a division of Madison Newspapers, Inc., in Madison, Wisconsin, holds seminars for local advertisers bringing in outside speakers to discuss creative ways of using cable. And in their newsletter "Cable Carrier," they even offered Free copies of this book to readers who phoned in!

When Clearview Cable in Tallahassee, Florida, began selling ads on CNN and ESPN in early 1981, they found it difficult to convince local businesses of cable's effectiveness as an advertising medium. Advertising agencies wanted documented numbers, which were not available, and Clearview even met resistance when it offered its spots on a free trial basis. To fight this problem, Clearview ran promotions, using the eight advertisers they already had managed to sell. This got people used to seeing local ads on the two channels. Over a period of a year, this combination of plugging spots, keeping rates competitive with radio, and just building overall awareness helped turn sales around.

And, finally, there was the rather unique approach used by the small (5,000 subscriber) Westerly, Rhode Island, system of Colony Communications. Going straight to local retailers, they sold a package of 10 spots a week for 52 weeks for $3,200 to the first 15 advertisers who signed up. At an average of $6 each, the 520 spots competed with local radio. Colony's target was local radio advertisers, and they succeeded, signing up three restaurants, two auto dealers, two box office supply shops, a jeweler, an appliance store, a shoe store, a florist, a home video store, a plumbing supply company, a grocery and an insurance agent.

The Local Commercial Production Problem—One Solution— A major stumbling block on the road to attracting cable advertisers who have not used television is the production of commercials. Palmer Cablevision, a cable advertising pioneer that operates systems in Florida and California, produced a package or 10 "generic" commercials for an appliance store, a jewelry store, an air conditioning/heating service, a women's and men's clothing store, a travel agency, an automobile repair shop, a furniture store, an insurance agency, and a real estate agent.

Palmer did market research to come up with the 10 different kinds of businesses known to be good broadcast advertising prospects. It then interviewed people in these businesses to find out their principal selling point. To spotlight the copy that was universally used in the various categories, they looked at the Yellow Pages.

Each 30-second spot contains 25 seconds of general audio and video content that relates to the specific business. Five seconds are reserved for the local cable system to insert the advertiser's tag with a camera card, a character generator, or a slide. This is the same concept that is found in the newspaper business with the production of local advertising mats.

Palmer markets a reel of the 10 assorted commercials for a total cost of $250 (only $25 a commercial) plus a tape charge. The advertising tapes appeal both to the smaller systems that do not have facilities to produce their own advertising messages *and* to larger systems that want to hold down advertising costs to appeal to small businesses that might not otherwise purchase cable.

Creative Associates of Louisville offers cable operators a similar service, but with a slightly different twist. They provide a reel of some 50 generic video spots free of charge to cable systems. The systems then show them to local advertisers who in turn buy the spots directly from Creative Associates, which customizes them with a "squeezoom" special effects generator or a simple voice-over. Advertisers provide copy and a camera card, and the spot is returned to them in seven days. Chuck Conrad, President of Creative Associates, says of his service: "I'd like to turn this into a giant production mail order business."

How Madison Takes the Worry Out of Production

At the local level, the production of commercials is more than just an "added service" of the cable system. It may be a necessity to attract marketers who have had no experience in developing advertising for television.

Madison Cable Network developed a simple production technique based upon following a specific format and scheduling shooting only on days when several clients can be serviced to gain economies of scale.

A professional crew is dispatched to videotape three or four scenes of raw footage for each client on the production schedule that day. The raw tape is then edited into a standardized 30-second message including the advertisers logo, address, and/or phone number. Next, a professionally produced sound track is added with the copy provided by MCN's creative department in consultation with the advertiser. The sound track can be inexpensively changed as often as once a week. And the advertiser can also add short printed messages to the spot (such as special promotion prices) with an electronic character generator.

Madison Cable Network's ability to take the local advertiser's mind off of the high costs and problems normally associated with commercial production and concentrate on the values of cable as an effective marketing medium is obviously a very effective sales tool.

Cable Salesmanship

Unfortunately, many marketers to whom a local cable system wants to sell advertising have little actual exposure to the medium itself. A local business owner watches television, reads magazines and newspapers, listens to the radio, and sees billboards day in and day out. He may hear a lot of people talk about cable—and he may even subscribe himself—but it's still something very new and untried. This makes the job of the local system operator even tougher when it comes to selling advertising to this prospective client. The local system must recognize that it is selling an entirely new medium. How well cable works for the advertiser will depend upon the degree to which the cable system operator understands the advertiser's needs and is able to deliver what will best satisfy those needs. Nowhere is the art of good salesmanship more necessary than in the selling of cable at the local level today.

To distinguish cable advertising from other local advertising opportunities, some system operators are creating unique "packages." Cable ad "avails" are offered in combination with spots in cable guides and direct mail tie-ins. The key is that the entire package is a one-price buy. (See Figure 9-1.)

In *Cable Advertising: The First Comprehensive Guidebook for System Operators*, Texscan Corporation highlighted the six key points of the local cable sales presentation.

1 Listen to his needs. Keep him talking about his product or service. The more he expands on his needs, the better your

Figure 9-1 The Cable Package

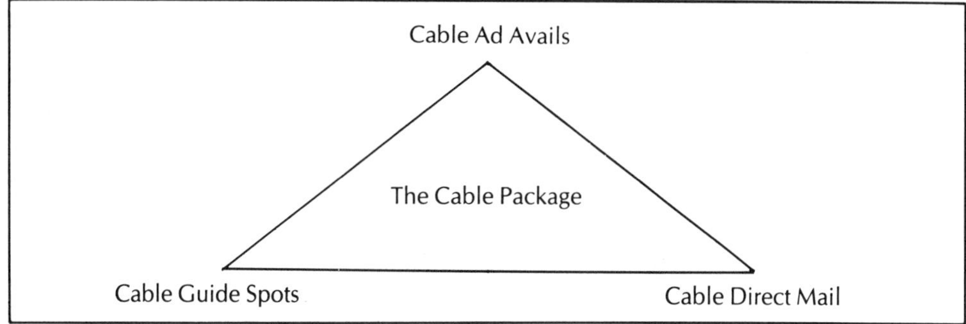

chances of supplying the creative idea that will sell him on cable.

2 Stress the benefits. The general benefits of cable and any additional ones that are particular to your system are indispensable as selling tools. Tell him all the things cable can do for him. Educate him!

3 Use examples of what others have done. If the customer is still hesitant about using cable after you have told him what it can do for him, give him examples of successful local advertising. You might suggest that he call some present advertisers and let them do the selling for you. Word of mouth is still one of the most effective selling techniques.

4 Invite him to see your system. If the buyer has never seen cable, you have to show him what cable is. Otherwise he won't be inclined to buy.

5 Show him he's the customer. Make him feel confident that you can fill his needs with cable better than any other media could. Let him do it his way. If he has an idea for a commercial message, jump on it. That's your chance to close.

6 If you are close to a sale, you can usually tell. When he starts asking a lot of specific questions, it's time for the close. Get a commitment. Ask for the order. Try not to let him put you off until a later date. Stress the importance of getting into cable now while the industry is young and prices are low. If he insists on making the decision sometime in the future, tell him you'll call on him again. Then mention a date in a week, a

> month, or whatever seems appropriate. Leave something behind—a rate card, a brochure or program schedule so he doesn't forget you. (*Cable Advertising: The First Comprehensive Guidebook for Systems Operators*, p. 10.)

Cable television advertising representatives need to understand the product they are selling as well as the basic principles of selling a communication system. A local cable system could find it beneficial to utilize the services of a communications training firm such as The Executive Technique in Chicago. According to John Connellan, President of The Executive Technique, "In order for cable ad sales people to be successful presenters, they must:

1 Improve their verbal skills by organizing their content and speaking to their listener's point of view rather than their own.

2 Improve their vocal and physical skills.

3 Improve their visual skills by the use of gestures, images and effective visual aids."

Sometimes, the best way to entice a local merchant into cable is to suggest that he try some simple but unique promotion idea to see if anyone out there is really viewing. In Carpentersville, Illinois, a service station ran a commercial five times a night for only one week offering a gallon of windshield solvent free if a customer mentioned the spot. During that week, the owner gave away 400 gallons of the solvent.

Local Cable Advertising and Program Production Guidelines

If cable video is to develop to its full potential at the local market level, it will be because there is established a solid working relationship between the advertiser and the local cable system. This is reflected in the following "Local Cable Advertising and Program Production Guidelines," developed by Tom Greer, former Vice President and Creative Director of the New Media unit of J. Walter Thompson, USA, and now an independent cable producer, director, and consultant. Tom's involvement in cable extends from creating the idea to selling it and developing the execution. He recognizes that if cable advertising is to succeed, it will be because of the intertwining of efforts by the buyer and the seller of the medium.

I. Identify the Local Cable System

The first step to the successful use of cable television in any community is getting to know your local cable operator. Establish what potential opportunities for advertising exist on the system. Discuss your mutual needs and interests. An ever increasing number of cable systems are providing local advertising opportunities. They generally fall into three categories:

1 Local availabilities in the national satellite, advertising supported services. Generally, the local cable operator has several minutes of time per hour to sell in cable services such as the Cable News Network, Entertainment and Sports Programming Network, MTV, USA, and The Weather Channel. The programs provide opportunities to advertise to a narrow target audience interested in news, sports, health, or culture.

2 Local-originated programming. Local systems generally have one or more channels that carry a variety of locally produced programs specifically targeted to needs within the community. At present these include:

Community News
Community Sports
Consumer Shopping
Local Celebrations and Parades
Community Business Profiles
Local Documentaries
Catalog Channels
Interviews and Talk
Community News
Community Cooking
Senior Citizen Events
"How-To"

Advertising, as well as full-program sponsorships, is available on most of these local-origination channels. You also have an opportunity to produce programs yourself, if your message is communicated better in this manner. While satellite network programs provide targeting of audiences, locally produced programs add the ability to have a total program "brought to you by . . . (you the advertiser)."

In addition, your local image can be enchanced by providing programming of a genuine community interest or need.

3 Automated programming and digital information. Alphanumeric channels provide information on weather, stock market quotations, community billboards, news, and sports highlights, and they usually have provisions for classified-style advertisements. While live demonstrations on programs are not possible, announcements about services, products, special sales, and so on are possible at very attractive prices. The key here is the buildup of cumulative viewing. While only a small number of viewers will watch these channels at any one time, most cable viewers will check the local weather, stocks, sports, or community information data channels at some time during the viewing week.

Some cable systems currently do not offer advertising availabilities or local-origination channels. They either have not considered the possibilities of additional revenue or currently do not have the equipment or services that allow for local advertising production. Do not give up, however! Show an interest in both the creation of local programming and local advertising production. In the near future, most cable systems will begin to consider both local advertising and programming as they recognize it can be a significant new source of revenue. Be there first. Prepare for the future. Have your ideas ready. If you show interest now, the cable operator will knock on your door first when local advertising and program production begins.

II. Identify Local Production and Creative Talents

Most cable operators come into the industry from business and banking. While they may have the technical equipment to produce programming and commercials, they do not have the expertise in the creative and marketing areas. Look to the cable system's program director, if one exists, and also the local colleges and universities, the local newspaper and radio station, small local media shops and creative boutiques who have people who can help you create storyboards, scripts, and identify where you might begin with local creative. Local franchisees and dealers who sell goods and services for national clients will find assistance from their own national advertising agencies. Workshops conducted by the Cabletelevision Advertising Bureau, the National Cable Television Association, and local cable clubs can also be of great help. Finally, look for creative assistance from those involved in advertising production for radio and print. Remember to go to the cable operator with a good creative plan in mind. Hopefully, the cable operator will have the necessary technical plan to help you execute your predetermined concept.

III. Assess Community Needs

In order to effectively use cable TV in your community, you must first evaluate both your own and your community's needs. Based upon this evaluation, you might choose specific existing national or local program environments in which to place commercial messages, produce an informational program series involving your product or service, or underwrite a series of community cable TV events. The best use of cable TV will be different in each community, and only you and your local cable operator can really determine what specific format of message delivery is best for you.

IV. Development of Creative Concepts

If you decide to produce commercials for cable, they will look very different from what you are familiar with on broadcast TV. "Freedom of choice," based upon the large number of programs and services available to the cable viewer, dictates that commercials that don't educate, inform, or entertain the viewer probably will not be watched. Experience shows that viewers can easily switch to CNN, local weather, a sports update, stock reports, and than cut back to other programming when the commercials are over. The addition of the remote control cable converter allows the viewer to check back quickly to monitor the re-start of a selected program, virtually guaranteeing that while missing the commercial, he or she won't miss any of the show. "Zapping," the avoidance of watching commercials, is a game played seriously by those who work to see how many hours they can watch cable without seeing a commercial message.

The creation of commercials and programs for cable demands creative executions that are attractive to viewer perceptions and interests and become something that adds value to the viewer's life. Cable video can communicate in-depth the "lifestyle" benefits of your product or service that cannot be communicated in a 30-second message.

V. Media Buying, Mixing, and Promotion

Cable video is a medium like other media—TV, radio, newspapers, magazines, outdoor, penny shoppers, etc. It must be examined within the context of these other media. Anytime when considering cable video, you should ask yourself, "What is it about my product or service that is of importance and of interest that I have not effectively been able to communicate in the past?" This is the key to cable advertising. The effective promotion of programming, particularly locally produced programs, is essential to its success. Use existing print ads and radio commercials to support your cable programs by producing short tags that tell the community about what you are offering on cable. Tie-in local sales, give-a-ways, and personal appearances with your cable ads to extend their effectiveness. Whatever you do in cable production, do it regularly, as awareness in your community about your cable programs will be cumulative, with small audiences that build each week. Cable will only be effective if you do enough, on a regular basis, to promote and tie-in your cable effort with your other media and be aware of your community's interest and needs.

VI. What Should I Pay for Cable?

Basically, you can expect to buy local cable media time at prices anywhere from $1 to $100 per advertising spot. This varies widely, and prices will also depend on whether you are talking about a 30-second announcement or a two-minute execution. If you buy a large enough schedule, many cable systems will provide production and editing at nominal charges or even for free. Negotiation with the operator of your cable system is important, and what you end up paying will be an amount that you both find acceptable. There are few rules in cable advertising and programming concerning costs, promotional trade-offs, etc. The only true rule in cable video advertising and programming is that good deals are not hard to come by. As in any media venture, potential for a long-term commitment will make a cable system operator more interested in you. If you are a long-term advertiser, you get the first opportunities for new ventures. When a cable system has programming ideas, he will approach you first, not your competitor.

CHAPTER 10

The Cable Tool Kit

It is perhaps not ironic that George Orwell's prophetic year "1984" arrived with the announcement of a new interactive innovation in television that could thrust the TV industry, consumers, and advertising into the future much sooner than most forecasters ever imagined.

The system—named ACTV—was developed by Catalyst Technologies, a company headed by Nolan Bushnell, founder of Atari. It is based on a "smart black box" attached to a viewer's television set and on new methods of program production. As examples of its potential:

- Viewers would see one standard picture of a football telecast with three other views of the same play action on the bottom of the screen. With a small, hand-held remote, they could switch back and forth between shots.
- In a news show, a viewer could set up his or her own broadcast in the order of his or her own interests, selecting expanded coverage of some stories or jumping to others.
- Program concepts would involve interactive soap operas, game shows, dramas, and personal help shows. An exercise program would allow viewers to determine their own workout routines.

- As many as 16 different commercials could be transmitted in a single commercial break. Viewers would see only the message the black box selected based upon their demographic and psychographic characteristics and interests.
- A consumer goods marketer with a variety of household cleaning products could broadcast a commercial in which the viewer would be asked to identify his or her own specific cleaning problem. Then the spot the home received would be aimed at solving its own specific problem. If no choice was made, the "basic" commercial would be shown. Similar commercial operations would involve recipes from a food company, travel information, and investment options from a brokerage firm.

The exact form in which ACTV might emerge—as programming on existing cable networks and/or as a network itself—will depend on the results of considerable testing. One thing, however, is certain. By the time the decade of the eighties comes to a close, ACTV and all the other new electronic media will have dramatically changed the world of video communications. For the consumer it will be confusing, but exciting. For the advertiser, it will be a real challenge.

The purpose of all advertising is to communicate a sales message that will convince a real person to make a real purchase of a real product or service. The role of media is to transmit that message in the most efficient way, whether it be the town criers who once wandered up and down our streets or any of the new media in all of their forms, present and future. The key to the successful use of the "new" media is the same as for the "old": offer a good product or service, identify your audience, speak to that audience in ways that are honest and meaningful, and don't be afraid to try a new idea.

In many respects this philosophy brings our discussion of the new media full circle. As you can see, it has been virtually impossible for me to escape dealing in the old "tried and true" aphorisms, and perhaps that is most appropriate. For just as we continually seek new ways of expressing existing ideas, the new media represent powerful ways of communicating these same ideas.

Throughout this book I have discussed and described the virtually limitless number of ways that you might use and experiment with the many forms of the new media. The rest of this chapter is devoted to providing additional suggestions. Called the "Cable Tool Kit," it is, like any good tool kit, designed to be used.

The Ben Franklin Cable Option Evaluator

It is reported that Ben Franklin developed a very simple technique for evaluating just about any set of circumstances.

1. *First,* he divided a sheet of paper into two sides—the left side (pros) and the right side (cons).
2. *Second,* he listed all of the factors supporting a recommendation on the pro side and all of the factors opposing it on the con side.
3. Finally, he "weighed" the two sides determining how each pro matched up against each con.

Whichever side outweighed the other determined his course of action.

The Ben Franklin Cable Option Evaluator lets you use this great man's same technique in deciding whether or not to invest in a given new media opportunity.

CABLE IDEA PROS	CABLE IDEA CONS
1.________	1.________
2.________	2.________
3.________	3.________
4.________	4.________
5.________	5.________
6.________	6.________
7.________	7.________
8.________	8.________
Etc________	Etc.________

The Cable Comparograph

As with other media, the evaluation of cable opportunities should focus not just on ratings and CPMs, but on other criteria that, while

JWT CABLE COMPAROGRAPH

		Subscriber Base(mm)			Home Reach[2]			Target Reach[3]								
Network[1] Name	Unit Cost	Current	+6mo.	+12mo.	%	(000)	CPM	Per Home	(000)	CPM	Commercial[4] Enhancing Environment	Commercial[5] Enhancing Message Load	Commercial[6] Creative Potential	Viewer[7] Actions	Advertiser[8] Identity Extras	Other[9] Considerations
1																
2																
3																
4																
5																
6																
7																
8																
9																
10																

Footnotes

1) If a specific program is being considered, indicate it by name.
2) Indicate average audience, quarter or half hour, or cume; Nielsen or estimate.
3) Indicate Nielsen or estimate and define target; level should reflect base in home reach.
4) Programming acts as more than just a commercial carrier and relates directly to product and/or target audience interests.
5) Commercial minutes per hour/commercial minutes per break.
6) Unique commercial lengths, formats, etc.
7) Indications and appraisal of any viewer response to promotions or offers by network or advertisers.
8) Potential for billboards, product on camera, name on set, etc.
9) Comment on merchandising opportunities, ability to reference other media in commercial, tie-in with print, etc.

more difficult to quantify, are nevertheless important in the analysis process. These factors include program and commercial environment, viewer actions, subscriber growth, and opportunities for creative commercial executions and sponsor identity.

The Cable Comparograph was designed as a structure for evaluating these factors in a relatively simple, organized manner.

A number of major advertisers and advertising agencies have gathered some of the information you may want to include. Other material can be secured from the networks themselves. While some of the data referred to may not be available or applicable in all circumstances, the structure is a valuable and simple tool to use in evaluating and comparing the cable offerings with which you are presented.

The Local Cable Idea Starter Kit

Earlier in this book, I noted that, in developing ideas for the effective use of cable, we are limited only by our imagination. Nowhere is this more evident than in the creation of cable concepts at the local level.

This section includes 104 ideas for effective cable selling concepts covering 42 product and service categories at the local level. And the challenge is the "blank space" in which you can add your own creativity.

Air Conditioning and Heating Companies

Home Insulation and Energy Conservation
____
____
____

Airlines

Exercises for the Business Traveler
Great Restaurants Around the Country (World)
Getting Ready for Your Vacation
____
____
____

Appliance Stores	Household Safety Tips
	Energy Saving Tips
Art Galleries and Dealers	What to Look for in Buying Art
Automobile Dealers	Where to Go and How to Get There
	How to Shop for a Car
Automobile Supply Stores	Do-It-Yourself Video Car Manual
Bakeries	Entertaining for the Holidays
	Tips on Party Planning
Banks and Savings and Loans	Personal Finance
	The ABC's of Getting a Loan
	The ABC's of the IRA

Bedding Companies	Interpreting Your Dreams
Boats and Marine Equipment	Boating Conditions in Your Area Tips on Sailing
Book Stores	Book Review of the Week The Top 10 Books of the Week
Building Material	How to Build Almost Anything What to Keep in Your Garage
Burglar Alarms and Security Systems	Safeguarding Your Valuables Protecting Your Home Against Intruders

Clothing Stores	Community Fashion Shows
	The Latest Fashions
	Clothes for the Working Woman
	How to Coordinate and Care for Your Wardrobe
Cosmetics and Beauty Aids	Make-up and Beauty Hints
	Taking Care of Your Face, Feet and Hands
Dental and Medical Services	Exercises at Your Desk
	Foods for a Healthier You
	Taking Care of Your Mind and Body
	Kick the Smoking Habit
Drugstores	Organizing Your Medicine Cabinet
	Lists to Leave for the Babysitter
Educational Institutions	Going Back to School After 30!

Employment and Recruitment Agencies	How to Interview for a Job
	Assessing Your Strengths and Weaknesses
	Writing an Effective Resume
Financial Services	Tips on Keeping Tax Records
	Understanding the New Tax Laws
	Tax Deductions You May Have Overlooked
	The Stock Market Report
	Planning the Family Budget
Furniture Stores	Caring for Your Furniture
	Arranging Furniture in Your Home
	Interior Decorating on a Budget
Garden and Lawn Supplies	Planning Your Summer Garden
	Caring for Your Garden
	Loving Your Plants
	New and Different Salads

Grocery Stores	Holiday Food Ideas
	Meals on a Budget
	Recipes for the Working Woman
	How to Fix a Last-Minute Dinner
	Planning Your Shopping List
	Summer Picnic Meals
	Seasonal Food Specialties
Hardware Dealers	Complete Do-It-Yourself Manual
	How to Repair Almost Anything
	Organizing Your Kitchen
	Women in the Hardware Store
Health Food Stores	Eating for a Healthier Life
	Putting Nutrition in Your Diet
Hobby Shops	Things to Do on a Rainy Day
	Fun for the Family
	Hobbies in Your Town

Home Improvements and Remodeling	Remodeling on a Budget
	Redoing the Kitchen and Bath
Hotels and Motels	Spending the Weekend in Town
	Entertainment Guide of the Week
Insurance Agencies	How to Buy Insurance
	How Much Insurance is Enough
Luggage Stores	Packing the Most in the Least Space
	Matching Your Luggage to Your Travel Needs
Movers	Getting Ready to Move
	Last Minute Moving Check-List

Office Equipment	Buying a Personal Computer
	Organizing Your Workspace More Efficiently
Pest Control and Exterminators	The Warning Signs of Pest Danger
	When to Call the Exterminator
Photo Equipment Stores	Photography Made Easy
	How to Photograph Children
	Choosing the Right Camera (Film)
Real Estate Firms	Video Home Tours
	Round-the-Clock Cable Home Listings
Restaurants	Favorite Meals of the Chef
	Favorites of the Celebrities

Service Stations	Getting Your Car Set for Winter (or Summer)
	Safe Driving Tips
	When Your Car Won't Start
Sporting Goods Stores	High School Sports
	Tips for (Golf, Tennis, etc.)
	Sportswear Fashions
Stereos and Hi-Fi's	Video Music Show (featuring local talent)
	How to Shop for a Stereo
Theaters	Entertainment Gossip News
	Movie Trivia Quiz
Travel Agencies	Vacations on a Budget
	Where to Go for a Weekend
	Exotic Spots to Visit

Veterinarians

Taking Care of Your Dog or Cat ______

Taking Your Pet on Vacation ______

Training Your Dog or Cat ______

______

______

______

Satellite Network Buying Checklist for

______

(Satellite Program Network)

Description of Satellite Network Programming

Overall: ______

______

______

Specific Program: ______

______

______

Basic Programming Appeal

Single Subject (Vertical) ______ or Multi Subject (Horizontal) ______

Loyal Audience (Low Turnover) ______ or Diverse Audience (High Turnover) ______

Merchandising Potential

To Sales Force, Dealers, Retailers, etc.: ______

______

______

To Viewing Audience: ______

______

______

Programming Schedule
Hours a Day____
Days and Times:____________________________________
____
Program Schedule (attached) Yes______ No______

Subscribers and Growth Potential
Satellite Carrier:___________________________________
Subscribers
Current:____
Year from Now:____________________________________
2 Years from Now:__________________________________
5 Years from Now:__________________________________
Subscribers by Market (attached) Yes______ No______
Subscribers by County Size (% of Total): A______ B______ C______
D______
Subscriber Characteristics (attached) Yes______ No______

Program Clearance Policies
Systems Carry Full Schedule____ Selected Programs/Segments____
Explanation (attached) if Systems Carry
Selected Programs/Segments Yes______ No______

Advertising
Commercial Minutes Per Hour: Network______ Local______
Nonstandard Commercial lengths______
Short Form (Under 30 seconds)______
Long Form (90 seconds)______ (Two minutes or more)______
Costs Per Commercial (30 second base)
Rate Card______
Negotiated______
Amount of Inventory Sold (estimated %)______

Audience and Demography
Estimated Household Rating:
Overall______ High Estimate______ Low Estimate______
Basis of Estimates:__________________________________
____
____
Viewer Demographics (attached) Yes______ No______

Maximum Schedule Reach Estimates
1 Day__________
7 Days__________
30 Days__________
Network Plans for Research__________

Network Agreement to Conduct Audience Research as Part of Advertiser Buy__________

66 Network Cable Idea Generators

The generation of an idea can begin from scratch. Or it can be built from an existing idea. Here are 66 of the most effective ways of creatively using cable which have been developed and implemented by advertising agencies and the cable satellite networks. Space has been left under each idea for your own creative concepts.

Ralston-Purina and Metropolitan Cats (ARTS)

Metropolitan Cats, a half-hour program produced in association with the Metropolitan Museum of Art explored everything from how we see cats in everyday life to how artists have depicted them in paintings from ancient Egypt to the twentieth century. Ralston-Purina recognized a good and unique fit between the subject matter of the program and their product and placed a variety of cat food commercials throughout the half-hour series. The synergistic blend of cats in the program and cats in the commercials demonstrated a creative capability of cable.

Promoting PBS on Cable (ARTS)

As a large underwriter of PBS' National Geographic Specials, Gulf always seeks ways to promote its sponsorship of the series. ARTS programming provided the opportunity to promote them to a similar upscale audience as PBS and, at the same time, deliver a traditional advertising message that could not be run on PBS.

Kraft "Family Favorite Recipe Contest" (BET)

Kraft sponsored BET's Family Favorite Recipe Contest with viewers invited to send in their family's favorite recipe using any Kraft product as an ingredient. Kraft judges selected the winners with the two first-place finalists winning a trip to Washington, D.C., where they made special guest appearances on Karen's Kitchen and prepared their winning meals.

Mobil Oil Player of the Week (BET)

Mobil Oil Player of the Week airs during half-time of each BET football and basketball telecast and salutes the most valuable player from the previous week's game. The segment features clips of the athlete's performance, comments on his overall statistics and positive information about his educational/career aspirations.

Minority Business Profile (BET)

Minority Business Profile (MBP) is a two-minute news segment sponsored by the Ford Motor Company, which reports on economic and business developments and on personal business success stories of primary interest to the minority community.

BET's Front Page (BET)

BET's "Front Page," a two-minute news update capsulizes news events that are of particular interest to Black Americans and was sponsored by American Telephone and Telegraph (AT&T) and International Business Machines (IBM).

United Technologies Commercial/Editorials (Business Times)

United Technologies Corporation ran a schedule of two-minute commercials which in many ways replicated Business Times' editorial product. When you saw their commercials air, you first thought it was part of the editorial format of the program.

Hourly Identification for Consolidated Freightways (Business Times)

Consolidated Freightways, Inc. produced 10-second corporate commercials, running one each hour for 52 weeks. They were positioned adjacent to a weather feature on Business Times, a subject of importance to a trucking company like Consolidated.

The First Video Tombstone (Business Times)

Morgan Stanley & Company, one of the largest brokerage houses in the country, used Business Times in a very unusual nontraditional manner. They had been trying to do something creatively for tombstones advertising for years. To develop and run a tombstone ad on television, however, was cost prohibitive. On Business Times, they created an inexpensive 45-second tombstone, the first ever to appear on the air.

Make It Easy, Make It Microwave (CBN)

A 13-week "how-to" exploration of microwave cooking techniques, Make It Easy was developed by J.C. Penney whose involvement went beyond commercial sponsorship. The program's hostess, Hilary Guild, is a home economist whose food preparation company was signed by Penney to conduct cooking demonstrations in their stores around the country. The microwave ovens were supplied by Penney, and Penney also helped select three co-sponsors—Corning Glass, Raytheon, and Rubbermaid—who each represented significant sales volume in their stores.

Selling Records for K-Tel (CBN)

K-Tel has advertised various record labels on CBN and the strength of the schedule is used to gain distribution on the album being promoted. CBN prints a sticker, "As Advertised on the CBN Cable Network," that is attached to each album in distribution. CBN also provides a program schedule that is sent out to each store and serves as a point or purchase promotion piece.

Swift Holiday Dinner Campaign (CBN)

Swift & Company produced a 30-minute program that creatively shows the viewer how to plan a holiday dinner. The dinner "planning" concept covers a three-or four-day period, preparing dishes daily so there is no last-minute rush on the holiday itself. The "star" of the show is Swift's Butterball Turkey. A two-minute infomercial airs simultaneously, promoting Swift products in recipes and a scatter plan of 30-second commercials is the final overlay, increasing reach and frequency.

The Amway "Teleconference" (CBN)

Amway has produced several commercial 30-minute programs highlighting the Amway Corporation and how to be an effective distributor. The programs are scheduled in order to reach all time zones from New York to Los Angeles and Amway notifies its distributors as to the air date and time so they can each have an effective teleconference.

The Disney Channel and the Family (CBN)

The Disney Channel used CBN as a commercial vehicle promoting their all-family program service. The Family Entertainer and Disney made an obvious team due to the similarity of program content.

The World of Amazing Animals (CBN)

Kal-Kan aired a 90-second feature on CBN entitled "The World of Amazing Animals." The program advised the viewer on pet care and a 30-second commercial was attached to it promoting various Kal-Kan products.

TWA/Kodak and Travelers World (CBN)

TWA and Kodak co-sponsored a CBN program, Travelers World, on which each week a different tourist location was spotlighted.

Cable Healthbreaks (CHN)

Like broadcast television "newsbreaks" in concept, the Cable Health Network's Cable Healthbreaks offered advertisers a 30-second island commercial within 60-seconds of programming directly related to the product or service being advertised. For example, for Aim Toothpaste, the subject matter is how to get kids to brush properly; for Menley-James (Contac), it is the causes and treatments of allergies.

Informathons (CHN)

A four-hour live program devoted to a single health issue. The first Informathon on heart disease was telecast in November 1982 and sponsored solely by Pfizer, Inc., a worldwide leader in health care products and services. As part of the full exposition on all aspects of heart disease, viewers could talk directly to heart specialists via a toll-free telephone number. During Heartline '83 in December 1983, over 100,000 phone calls were received from viewers around the U.S. The informathons are now on Lifetime.

Physician's Journal Update (CHN)

A two-hour program aimed strictly at keeping doctors and health care professionals up-to-date on the latest developments in the medical field, Physician's Journal Update discussed (in technical terms) new articles in major medical and scientific journals. It also provided pharmaceutical manufacturers (Ciba-Geigy, Parke-Davis, Pfizer, Searle, Squibb, etc.) with the first-time-ever opportunity to advertise prescription drugs to their primary target market (physicians) via television.

A Franchise in Business (CNN)

Inside Business, a 30-minute program created for Lanier Business Machines, featured Myron Kandel interviewing CEO's of major corporations. Lanier liked the concept and the sponsorship so well they committed to a five-year contract.

Juice, Cereal and Nutrition (CNN)

CNN's four-six minute nutrition segments were created for sponsorship by the Department of Citrus and Total Cereal. The vignettes ran several times per day and each day included a fresh vignette.

Air Travel and Travel Lodge (CNN)

Air Travel Advisory vignettes were created for Travel Lodge. The purpose is to give air travelers a fast update of traffic delays at major airports throughout the country and the information is received directly from the Civil Aeronautics Board.

The Infomercial-Commercial Connection (CNN), (WTBS)

Swift runs two-minute infomercials—small stories on how to use Swift products in a variety of exciting, new ways—followed within 20 minutes by a 30-second commercial for the same product.

Getting Your Opinion Across (CNN)

W.R. Grace used CNN for some very fast advocacy messages on a political issue the company felt was important.

Parents and Pictures of Kids (Daytime)

Sears Portrait Studios is primarily a photographic service for young children. By running spots within the parenting segments in Daytime, Sears would not attract an audience as large as for network television, but the vast majority of its viewers would be mothers of small children.

Women and Auto Maintenance (Daytime)

Lucille's Car Clinic let Mobil speak to women on the subject of auto maintenance. While Mobil's normal target audience was men, this gave Mobil the opportunity to reach women for its Mobil I motor oil. Shown on location in her transmission shop, Lucille demonstrated how to diagnose and repair auto problems. Each program included active audiences participation and a tip such as how to use your pantyhose or necktie to replace a broken fan belt.

"Do You Know Me?" (ESPN)

Tying a show concept in with their popular advertising campaign concept, American Express sponsors a regular ESPN program segment featuring outstanding sports figures.

Close Shaves by Noxema (ESPN)

Relating a product benefit to a program idea, this ESPN feature highlighted close sporting event victories.

Selling Sports Store Franchises (ESPN)

Sports Shack, a $60,000 turnkey sports store franchiser, was looking for leads. Running only 15 30-second units, they received 600 replies that turned into 86 very qualified leads.

Auctioning a Painting (ESPN)

NYC Art Store auctioned a $120,000 Norman Rockwell original oil painting using 14 prime-time one-minute spots over two weeks. They received 60 bids of which a dozen were deemed qualified and the painting sold for $135,000.

A Timely Buy by Timex (ESPN)

Timex made a 10-year deal to have their clocks be the official timepiece of ESPN. Every time a clock is shown in any ESPN sporting event, it is a Timex.

Long-Term Beer Exclusivity (ESPN)

Anheuser Busch made a five year $25,000,000 commitment to get total beer exclusivity on the network. With long-term price protection, the buy became even better as the years went on.

The Financial News Network Cross-Plug (FNN)

American Airlines was the first major advertiser to use FNN's cross-promotion campaign. The network produced (at no charge) 10 to 20-second tags which followed the airline's commercials. The tags identified the airline's sponsorship of sports programming and specials on other services and broadcast outlets and encouraged viewers to tune them in.

The American Airlines Connection (FNN)

The Financial News Network produces two 15-minute business and financial reports which air on all East and Westbound American flights of three or more hours. Sponsorship allows companies to reach a very upscale market—the airline traveller who is also keenly interested in the world of finance and business.

Kraft Goes New Wave (MTV)

Kraft produced two special new wave spots for airing on MTV. The cost for both commercials was under $25,000 and without jeopardizing their traditional, loyal audience, Kraft was able to look "hip" and reach a new customer base via MTV.

Chevrolet in Stereo (MTV)

To introduce the new Camaro, Chevrolet created a 90-second stereo spot for MTV. The commercial was produced in such a manner as to allow one 60 and two 30's to be cut for use on over-the-air broadcast networks. This let Chevrolet reach MTV's highly targeted audience with little additional cost and with a stereo rock soundtrack that tied in with MTV's programming.

Jensen Car Stereos Enter TV (MTV)

Jensen, a small, national print advertiser, produced their first-ever TV spot for MTV only. Because the audience *was* their customers, the spot was "new wave," one that might have had less than optimistic reception on more traditional TV.

Raisinets Goes to the Movies (MTV)

A brand uniquely associated with movies in the consumer's mind bought MTV in a special fashion. A 10-second ID was run prior to movie commercials on MTV. The voiceover was "Now, Raisinets brings you this preview of an exciting new motion picture," thus implying sponsorship. The ID was lifted from the already produced 30-second spot, so little additonal cost was involved.

Welcome to Miller Time (MTV)

Miller took their well-known jingle, "Welcome to Miller Time," and had a rock group (Southside Johnny) sing it in a music club setting. The spot was tagged in an identical fashion to MTV music video clips, thus causing the viewer to think he was watching another music clip. By the time the realization that a commercial was being played occurred, the viewer (hopefully) was humming the jingle.

Recruiting a Sales Force on Cable (MSN)

The World Book Encyclopedia used an infomercial on The Home Shopping Show to recruit people for its sales force who work on a commission basis and set their own work hours.

Selling Gadgets Direct on Cable (MSN)

The Sharper Image uses the Modern Satellite Network to sell their expensive toys and gadgets. An 800 toll-free telephone number allows the at-home shopper to buy direct.

Addressing the Country on Cable (MSN)

Herb Schmertz, Vice President of Public Affairs for Mobil Oil Corp., addressed a convention of public relations professionals meeting in Florida. He was "live" in New York and spoke to the delegates in Florida via MSN on First Amendment rights.

Saran Wrap, Rubbermaid and Microwaves (SPN)

On the Satellite Program Network's Microwaves are for Cooking, both Saran Wrap and Rubbermaid demonstrated the uses of their products in conjunction with home microwaves.

The Ralston Reward (SPN)

On the Satellite Program Network's program Companion Dog Training, Ralston-Purina products were used as a reward when a dog performed a command well.

A Country "Package" for Wrangler Boots (TNN)

New to television advertising, Wrangler, in association with The Nashville Network, produced three 30-second commercials using singer/songwriter Ed Bruce as spokesperson. Special segments on Nashville Now featured the Wrangler Boot factory. In addition, Wrangler Boot gift certificates were given out on the program. Finally, the Wrangler Star Search, a talent contest aired on TNN.

Promoting Kraft and Country (TNN)

To add extra mileage to the Kraft Nashville Network schedule, special "tags" were produced for the week prior to the Kraft-sponsored Country Music Association Awards telecast on CBS. Each night a CMA nominee appeared on Nashville Now and Ralph Emery announced the nominees in the category (:20). The Kraft commercial followed (:30). A tag (:10) followed the commercial promoting Kraft's involvement with the CMA and alerting viewers to watch the show on CBS and listen to the simulcast on Mutual Radio.

Ken-L-Ration and Country (TNN)

Quaker's schedule on The Nashville Network was supplemented with a special pet segment featuring Tom T. Hall and his wife Dixie. Dixie, a breeder of basset hounds, gave TNN viewers an inside look at her kennels and her award-winning hounds. The segment was fused to a live, on-air mention by Ralph Emery of Ken-L-Ration's Kibbles & Bits, and the Kibbles & Bits commercial.

Cat Chow and Celebrity Softball (TNN)

The Nashville Network aired the Barbara Mandrell/Conway Twitty Celebrity Softball Classic, which was played for the benefit of the National Humane Society. TNN produced promos that featured Barbara Mandrell speaking on behalf of the Humane Society and Ralston-Purina's ongoing support of this most worthy cause. The promos ran for several weeks prior to the airing, and Ralson's Cat Chow then sponsored the game.

Kellogg on the Set (TNN)

Kellogg, sponsoring I-40 Paradise on The Nashville Network, obtained added promotional consideration when it was arranged to have Kellogg's Special K cereal on the set of the cafe.

Atari on Radio 1990 (USA)

Atari purchased a franchise position in Radio 1990 to reach the program's teen/young adult market for video games. Their commercials are complimented with a "drop-in" element—Top 10 Albums of the Week—followed by a different video clip each day from one of the albums. The segment carries Atari video game graphics and billboards identifying it as an exclusive Atari property.

Alive & Well with Bristol-Myers (USA)

Bristol-Myers' 10-year, $40 million commitment to USA was a major advertiser investment in cable programming. Its participation in Alive & Well provides Bristol-Myers with 21 hours of women's health and fitness programming to exclusively showcase its many products and has spawned Alive & Well magazine and Alive & Well books. Direct response ads for diet booklets and magazines containing redeemable coupons have produced throusands of diet book orders and magazine orders.

Armstrong and Better Homes & Gardens (USA)

Armstrong World Industries sponsors Designs for Living with Better Homes & Gardens, whose environment matches their home improvement and design products. The show (and by extension, Armstrong) also receives extensive cross-promotion in Better Homes & Gardens.

Mazda Looks at Sports (USA)

Mazda Motor's sponsorship of Sports Look was among the first advertiser commitments to cable programming. The sports interview show, produced by their agency, Foote, Cone & Belding, provides Mazda with both an effective marketing vehicle and a highly targeted audience.

Quaker Data and The Weather (TWC)

To capitalize on the relationship between cold weather and hot cereal sales, Quaker runs adjacencies to Weather Channel local forecasts during the cold weather months. Research shows that sales of hot cereals peak when the daytime temperature is 40 degrees or less. When The Weather Channel forecasts this condition in 15 or more states, a Quaker Cold Weather Alert is called, and Quaker's spot frequency is increased from 6 occasions per day to 16 per day for a two-day period.

Commodities, The Weather and Prudential-Bache (TWC)

To reach the high-tax bracket investor who needs advice on commodities, The Weather Channel, in conjunction with Prudential-Bache, developed the Agricultural Weather Report. It covered subjects of interest to commodities traders, ranging from current rainfall and harvesting conditions and the effect of severe weather on livestock to the longer term effect of weather on anticipated crop yields.

Pest Control and High Target Marketing (TWC)

To capitalize on the relationship between pest control and the weather, Orkin developed a series of tips using Weather Channel talent to inform viewers about the need and ways to control pests. The tips were followed by an Orkin commercial, after which specifically localized information was inserted automatically on a cable-system-by-cable-system basis by the Weather STAR.

When It Rains... It's Anco Time (TWC)

In a match-up of product usage and the advertising environment, Anco Windshield Products sponsored national precipitation forecasts on The Weather Channel plus adjacencies to local weather forecasts. Anco also uses The Weather Channel's ability to deliver localized alphanumeric messages on a cable system-by-system basis to direct Weather Channel viewers to their local Anco dealer.

The Bounce Static Report (TWC)

Bounce fabric softener's selling proposition includes its antistatic cling properties. When indoor heating dries the air, static cling results from the lowered indoor relative humidity. The Bounce Static Report takes viewers through a series of national maps showing temperature, outdoor relative humidity, and indoor relative humidity. The meterologist also makes reference to problems caused by lower indoor relative humidity, which include dry skin, loosening of furniture joints and static cling.

Weatherizing Your Pets(TWC)

Pets can be affected by weather conditions, and The Weather Channel developed a series of 30-second pet tips to provide viewers with information about how to Weatherize Your Pet. Designed for each season, these mini-features run with 30-second Ralston-Purina commercials.

Flying the Friendly Skies. . . City by City (TWC)

Both business and pleasure travelers are vitally interested in the weather forecast for their destination, so United sponsors The Weather Channel's city-by-city forecast, which covers major business traveler destination cities, the Vacationer's Atlas covering the outlook for vacation regions around the country and The Hawaii Briefing. United uses The Weather Channel's ability to localize by showing fares and schedules appropriate for each market, with different data for each program they sponsor. The highly competitive and sometimes volatile nature of the market means that fares and schedules can and often are changed with 24-hour notice.

A Soap Opera for Cable (WTBS)

Proctor & Gamble co-produced and sponsored The Catlins, a new serial airing mornings and late night on WTBS. The primary motivation was control over the content and commercial inventory of a program—and attractive cost efficiences.

The Coors Sports Page (WTBS)

Without national distribution, most major live sports events are neither practical nor economically feasible for Coors. The weekly Coors Sports Page on WTBS allowed the brewer to create a long-term franchise against their target at an affordable level.

Warner Brothers at the Movies (WTBS)

Warner Brothers sponsors vignettes promoting their youth-oriented movies. Airing Friday and Saturday nights in Night Tracks, they reach a youth audience of heavy movie-goers.

The Bristol-Myers Theatre (WTBS)

On the first Monday of each month, Bristol-Myers sponsors the WTBS prime-time movie. The specific film shown is based upon a monthly poll where viewers select from a list of several options what movie they would like to see.

Nutrition in the News (WTBS)

G. D. Searle sponsors nutrition vignettes in WTBS' prime-time news to provide their sugar substitute, Equal, with an appropriate programming environment that can also efficiently deliver their target.

Fly Eastern to Atlanta (WTBS)

A consumer-oriented merchandising effort tied to a scatter plan, this approach gives viewers the chance to win a two-for-one flight to Atlanta (home of WTBS) with all room, board, and entertainment expenses paid for by Eastern and WTBS.

Sources of Additional Information

Below are listed some standard sources of information about the new electronic media. Since the number of publications and organizations devoted to aspects of the cable industry has increased manyfold during the past few years, any list will be incomplete. This list should, however, provide a good starting point.

A.C. Nielsen Company. New York.

Advertising Age. Chicago. Weekly.

Arbitron Company. New York

Audits of Great Britain. London/New York.

Broadcasting. Washington, D.C. Weekly.

Cable Marketing. New York. Monthly.

CableAge. New York. Weekly.

CableFile. Denver. Annual.

Cabletelevision Advertising Bureau. New York.

CableVision. Denver. Weekly.

Editor & Publisher. New York. Weekly.

Electronic Media. Chicago. Weekly.

Information & Analysis. Hicksville, N.Y.

Information Resources Inc. Chicago.

Marketing & Media Decisions. New York. Monthly.

Marquist Media Services. Beaufort, N.C.
MediaMark Research Inc. New York.
Multichannel News. Denver. Weekly.
National Cable Television Association. Washington, D.C.
The New York Times. New York. Daily.
NPD Group. Port Washington, N.Y.
Simmons Market Research Bureau, Inc. New York.
Statistical Research Inc. Westfield, N.J.
Television Audience Assessment. Cambridge, Mass.
Television Digest, Inc. Washington, D.C.
Television/Radio Age. New York. Biweekly.
TVC. Englewood, Colorado. Biweekly.
View. New York. Monthly.

A Glossary of 123 Essential New Media Advertising Terms

Access Cable. Local channel space available either free or on a leased basis to individuals and groups in the community, including educational and governmental bodies.

Adjacency. A local or spot commercial that runs before or after rather than within the main body of a show.

Advertorial. An advertising message that is longer in length than the typical broadcast commercial message and provides in-depth information on a product, service, or company. (*See* Infomercial).

Alphanumeric. Information such as news, weather, sports, stock quotations, and advertising that is projected on the screen in numerical and letter form via a character generator.

Arbitron. A research firm that gathers radio and television audience data and conducts special audience studies for cable systems and satellite networks.

Area of Cable Influence (ACI). The geographic area served by a cable company. It may include an entire town or only a portion of it in the case of most large cities.

Area of Dominant Influence (ADI). Arbitron's geographic market definition that includes all counties in which its television stations achieve their greatest audience. An ADI may include many different *Areas of Cable Influence (ACI).*

Audience. The number or percentage of people or homes exposed to a program or other advertising vehicle.

Audience Composition. The distribution of a program's audience by age, sex, income, education, and other categories.

Audience Duplication. The number or percentage of viewers of one program who also are exposed to another program or advertising vehicle.

Audience Profile. The demographic characteristics of the people or households viewing a program, station, cable system, or other advertising vehicle.

Availability (Avails). The commercial time that a cable system, station, or network has for sale.

Average Audience (AA). A Nielsen term that reflects the number of households or persons viewing the "average minute" of a program's duration.

Basic Cable. Programming offered to a cable system's subscribers at a "basic" monthly fee.

Billboard. Usually a 3 to 10 second announcement at the beginning or end of a show that identifies the sponsor(s) of or advertiser(s) in the show.

Bonus Spot. A commercial given to an advertiser at no charge to make up for the deficiency in audience delivered by his schedule or as an incentive to get him to buy other spots.

Break. The time between or within a program used for commercials, announcements, or newsbreaks.

Broadcasting. The transmission of radio and television programs by over-the-air signals.

Cablecasting. Cable programming fed from a cable system to homes via coaxial cable rather than by over-the-air signals.

Cable Catalog. A direct response cable advertising/sales service where products are shown in catalog form on the TV screen and where viewers can make purchases via an 800 number.

Cable Penetration. The proportion of all television homes in an area that subscribe to cable. If there are 100,000 television homes in a market and 20,000 cable subscribers, the penetration level is 20 percent.

Cable Television. The transmission of television programs available from a master antenna, additional programming services distributed by satellite, and programming originated locally at a cable system to subscriber homes by cable (wires) instead of over-the-air.

Channel Capacity. The number of different channels of programming or services, some of them advertiser supported, available to subscribers of a cable system.

Character Generator. A typewriter-like device that projects information onto a television screen for the transmission of news, weather, sports, financial data, and alphanumeric advertising.

Chroma Key. A videotape effect in which one image can appear against one or more different backgrounds

Churn. The turnover in subscribers to a cable system due to new sign-ups, cancellations, etc.

Circulation. The number of different persons or households that tune in a broadcast or cable signal during a specified period of time.

Coincidental Interview. An audience survey in which people are asked what they were viewing at the moment they were contacted.

Commercial Protection. The agreement of a network, a station, or a cable system not to schedule a competitive product or service within a certain amount of time from an advertiser.

Communications Satellite. A vehicle located 22,300 miles in space that receives and transmits a variety of network news, sports, cultural, ethnic, entertainment, and other programming to cable systems around the country.

Community Antenna Television (CATV). The early name for cable television, which referred to the distribution of television signals via a master antenna to homes that either could not receive them with their own antennaes or received them very poorly.

Composite Master. The completed videotape into which all elements have been edited.

Cost Per Point. A media planning tool that represents an estimate of the cost of one rating point of commercial time in a particular daypart, program type, or overall buy.

Cost Per 1000 (CPM). Another advertising planning and evaluation tool that represents the cost of reaching 1000 homes or individuals. It is the most commonly used method of comparing cost efficiencies for different programs or schedules.

Coverage Area. A geographic area in which a broadcast or cable signal can be received.

Crawl. A list of names that moves across a video screen. It can be used to list the dealers in a market who carry a product advertised in a cable commercial.

Crop. The desired composition of a picture created by a camera.

Cumulative Audience (Reach). The number or percent of different people or homes reached by a schedule of programs, commercials, or advertisements over a specified period of time.

Cut. An editing technique to change one visual very quickly to another.

Dedicated Channel. A cable channel that is totally devoted to carrying a single source of programming.

Demographics. Audience composition or characteristics based on a variety of economic and social factors and used to evaluate programs.

Deregulation. Actions by the Congress and FCC to do away with or reduce restrictions involving the communications industry over which they have control.

Designated Market Area (DMA). Nielsen's geographic market definition that includes all counties in which its television stations achieve their greatest audience. Like Arbitron's ADI, it may include many different *Areas of Cable Influence.*

Diary. A research survey tool in which people record their viewing activity over a specified period of time. Either one *diary* is completed for each household or there is an individual diary completed for each viewer.

Direct Broadcast Satellite (DBS). A satellite transmission service that delivers a signal to a viewer's home directly via his *own* earth station and not through a cable system.

Direct Response Advertising. Commercials that use telephone numbers or addresses to permit viewers to order merchandise or get additional information on a product or service.

Downlink. That part of a satellite transmission system in which programming or other information is transmitted from the satellite to the ground.

Dub. To transcribe the sound or picture from one recording to another.

Earth Station. A dish-type antenna and other equipment that makes up a communications station used to send or receive programming and other information to or from a satellite.

Edit. To eliminate and join together portions of a film or tape by editing and splicing.

Electronic Newspaper. Newspaper-like textual information (news, sports, and stock listings) that may be accompanied by advertising and is transmitted on a video screen.

Fade In/Fade Out. An editing technique in which a subject appears or disappears slowly on the screen.

Fiber Optics. Thin glass fibers used to transmit light waves that, in turn, transmit information. Substantially more information may be transmitted via fiber optics than on radio waves.

Flighting. A manner of scheduling advertising so that it runs in-and-out over a period of time rather than continuously.

Fragmentation. A term that describes how the growing number of uses to which the television set can be placed (Cable, VCR, Videodisc, etc.) is resulting in a splintering of audience levels and declining broadcast television shares.

Franchise Area. The geographic area that a local government awards to a cable company and in which it provides cable service to subscribers.

Frequency. The average number of exposures received over a specified period of time by the different homes or people reached by a schedule of programs or commercials.

Gross Impressions. The sum total of the audiences of people or homes reached by each program or commercial in an advertiser's schedule of announcements.

Gross Rating Points (GRP). The same as gross impressions but expressed as the total of all rating points achieved over a period of time.

High Definition TV (HDTV). Television systems that are being experimented with and will provide sharper picture definition than the current U.S. standard, 525 lines per frame.

Homes Using Televison (HUT). The total available video audience at a given time as reflected by the number or percentage of television homes watching broadcast or cable programs. (More properly called Homes Using Video).

Infomercial on Informercial. A long-form (2-7 minutes or more) commercial message that provides in-depth, helpful information about a product, service, or company. (Also referred to as *Advertorial).*

Interactive Cable. A two-way system in which viewers respond to what is on the screen by pushing buttons and ordering merchandise, participating in viewer polls, or requesting information.

Interconnect. Several cable systems in a given area that join to sell commercials and offer the convenience of one order-one bill advertising placement.

Jingle. A commercial message's music and lyrics.

Local Origination Programming. Programming produced by a local cable system. It usually focuses on local events and community affairs and, in many cases, offers local sponsorship opportunities.

Location. A shooting site outside of the studio.

Make Good. A commercial offered as an alternative to one that did not run as scheduled and was lost by pre-emption, withholding or mechanical failure.

Media Plan. The description of various media that will be used to achieve an advertiser's marketing goals.

Meter. An electronic audience measurement device that automatically records broadcast or cable viewing.

Minicam. A small, portable color camera used to cover events and tape commercials outside of the studio.

Multiple System Operator (MSO). A large cable company that owns many systems throughout the country.

Multipoint Distribution System (MDS). A pay television service that transmits programming (largely movies with some specials and sports) via microwave for generally short distances to subscribers who pick them up with special antennae and converter boxes.

Narrowcasting. A cable program or program service that appeals to a "select" demographic target or special interest group rather to a "broad" segment of the population. A better term is perhaps Targetcasting.

Network. Broadcast stations or cable systems that are linked together by microwave, satellite, telephone lines or coaxial cable and receive programming delivered from a central distribution or production point.

Nielsen. A research firm that gathers television audience data and conducts special audience studies for cable systems and satellite networks.

Outtake. Film or taped footage that is not used in the final edited program or commercial.

Participation. A commercial that appears within the body of a program rather than between two different shows.

Pay Cable. A cable service that provides programming (largely movies, sports, and specials) for which a subscriber pays a charge in addition to his basic monthly fee.

Pay Per View Television. A pay television service for which the subscriber pays for each program viewed rather than per month.

Per Inquiry (PI). Direct response advertising in which the advertiser pays for each sale or response generated rather than for each commercial

Persons Using Television (PUT). The total number or percentage of people watching broadcast or cable televison programs at a particular time.

Pre-emption. A program that does not air but is replaced by another show or special event or a commercial that was scheduled to run but does not for any one of many reasons.

Quintile. The division of a group into equal fifths, such as cable viewing *quintiles*, from heaviest to lightest viewing *quintile*.

QUBE. Warner-Amex's two-way interactive cable service, which was developed in Columbus, Ohio, and has spread to other markets where Warner-Amex has won cable franchises.

Rating. An estimate of a broadcast or cable audience's size expressed as a percentage of all people or homes in a given demographic category.

Reach. (See **Cumulative Audience**).

Recall. A research technique in which a person is asked to remember what he saw, heard or did earlier in the day, week or longer ago.

Sample. The group of people selected to take part in a research study.

Satellite-Fed Master Antenna Television (SMATV). A minicable system for buildings connected to a private satellite antenna. It provides multichannel video to large apartment buildings and condominium complexes.

Saturation. A commercial buying technique where many announcements are run during a short period of time to reach a maximum number of homes or people.

Scatter Plan. A commercial buying technique where announcements are spread across many different programs or time periods rather than being confined to any particular one. An example would be running one commercial every four hours on ESPN over a week rather than concentrating them all in one or two football games.

Share of Audience. The percentage of all homes or people watching television at a given time who are watching a particular program, station, or channel. For example, if 28.0 percent of the homes in WTBS' coverage area have their TV sets on at 11:00 A.M. on Sunday morning and 7.0 percent are watching WTBS' Movie 17, it has a 25 percent share of audience.

Spill-in. The percentage or number of homes in one television market watching a station that originates in another market.

Spill-out. The percent or number of homes outside of a particular market watching a station located in that market. Since cable is transmitted to homes via wires rather than over-the-air, there is no spill-in or spill-out.

Splice. The combination of two or more different pieces of film into one continuous reel.

Sponsorship. The purchase by an advertiser of a sufficient amount of time in a broadcast or cable program to have product exclusivity, billboard identification, and, often, long-term price increase protection.

Spot. Commercial time available for sale by a broadcast station or cable system to local or national advertisers.

Still. A "nonmotion" photograph. Many cable systems will prepare commercials for local advertisers using stills and voice-over announcer copy.

Subscription Television (STV). A noncable pay television program service (largely movies with some sports and specials) in which a scrambled signal is sent out over-the-air and unscrambled by a decoder box on the viewer's television set.

Super. The superimposing of copy on a television screen. For example, a national automobile advertiser can provide commercial footage to a cable system and ask them to *super* the local dealer's name across the end of it.

Superstation. An independent television station (e.g., WGN in Chicago, WOR in New York, and WTBS in Atlanta) whose signal is transmitted via satellite and picked up by earth stations at cable systems across the country.

Tag. Information added locally to a commercial such as the name of a retailer where an advertiser's merchandise can be purchased. In another case, a food advertiser might take a commercial that has run on broadcast television, add a tag making available a free recipe booklet and run it in a women's oriented satellite cable programming.

Target Audience. Those persons defined in terms of their demographic characteristics or purchase behavior who are most desired by an advertiser because of their anticipated consumption of his product or service.

Targetcasting. A new term used to refer to cable programming that appeals to a "targeted" or "select" demographic audience in much the same way as Narrowcasting.

Teletext. The transmission of "pages" of textual information to a TV set, either over-the-air on a broadcast channel or by wire to a cable channel. With a specially adapted home television set, viewers may retrieve specific information that they desire such as news, weather, sports, finance, etc. Advertising messages and identification may accompany the transmission of the data.

Tiering. The different levels or groups of cable services available to subscribers at different monthly charges.

Time-Shifting. The use of video recorders by viewers to tape programs at one time and play them back at another, thus enabling them to see shows they otherwise would miss because they were not at home, were occupied with other tasks, or were viewing another program.

Total Audience. The percentage or number of homes or people who view some part of a program (usually at least six minutes). It is larger than the average audience and gives a general indication as to overall tune-in to the show.

Transponder. That part of a communications satellite which receives a signal from a specific program source and transmits it to earth stations at cable systems around the country.

Two-way Cable. The transmission of signals both ways along the cable that permits interactivity between viewer and cable system. (See **Interactive Cable**).

Universe. The total population or group being studied in a research project.

Uplink. That part of a satellite transmission system in which programming or other information is transmitted from the ground to the satellite.

Video Cassette Recorder (VCR). A machine that hooks into a television set and records programming on videotape for playback at a later date or plays pre-recorded tapes that are available to buy or rent.

Video Disc System. A machine that cannot record but that plays back pre-recorded discs available for sale.

Video Games. Electronic games that attach to the television set (and usually result in 10-year-old kids defeating 40-year-old adults by wide margins!).

Videotex. A two-way interactive cable or telephone provided service in which data stored in a central computer is retrieved on home screens through the use of home terminals. Videotex also permits home shopping and can offer advertising opportunities.

Viewers Per Viewing Household. The average number of people watching a program in each viewing household.

Voice-over. An announcer's voice that is heard in a commercial or program. For cable, an advertiser might take an existing broadcast commercial and add a different "voice-over" with a direct response offer.

Wide Screen. A large (up to six feet) screen television set with front or rear projection.

Wipe. The special visual effect where one object gradually replaces another across the screen.

Zapping. The conscious effort on the part of viewers to switch from one channel to another during commercial breaks and thereby avoid the advertising message(s).

Zoom. The camera effect whereby a subject is made to move closer in or farther away from the screen.

Index

a

e

f

g

i

j

k

o

p

q

t

U

y

z